The Anatomy of Body Worlds

The Anatomy of Body Worlds

Critical Essays on the Plastinated Cadavers of Gunther von Hagens

Edited by
T. Christine Jespersen,
Alicita Rodríguez
and Joseph Starr

McFarland & Company, Inc., Publishers
Jefferson, North Carolina, and London

LIBRARY OF CONGRESS CATALOGUING-IN-PUBLICATION DATA

The anatomy of Body Worlds : critical essays on the plastinated cadavers of Gunther von Hagens / edited by T. Christine Jespersen, Alicita Rodríguez and Joseph Starr.
p. cm.
Includes bibliographical references and index.

ISBN 978-0-7864-3656-9
softcover : 50# alkaline paper ∞

1. Hagens, Gunther von — Exhibitions. 2. Human anatomy — Exhibitions. 3. Cadaver in art — Exhibitions. 4.Human figure in art — Exhibitions. 5. Tissues— Plastic embedment — Exhibitions. 6. Plastination — Exhibitions. 7. Medicine and art — Exhibitions. I. Jespersen, T. Christine, 1964– II. Rodriguez, Alicita, 1972– III. Starr, Joseph, 1969–
QM16.H34A53 2009
612 — dc22 2008041143

British Library cataloguing data are available

Manufactured in the United States of America

McFarland & Company, Inc., Publishers
Box 611, Jefferson, North Carolina 28640
www.mcfarlandpub.com

Acknowledgments

We wish to thank Sally Sanchez, Phil Heldrich, and Ken Dvorak of the Southwest/Texas Popular Culture and American Culture Associations for hosting our panel presentation "Body Worlds: The Gender, Art, and Science of Plastinated Cadavers" and to the members of the audience who improved our papers through thoughtful conversation. Many thanks to the librarians at Western State College of Colorado's Savage Library, especially Nancy Gauss and Patrick Muckleroy, who answered questions and procured hundreds of articles and books for us. Thank you to the following organizations for helping us contact interested contributors: the University of Pennsylvania, the Radical Philosophy Association, and the Society for the Study of Symbolic Interaction. Thank you to Blackwell Publishing and Indiana University Press for permission to reprint essays by Uli Linke and Peter McIsaac. Our gratitude also goes to Anthony d'Offay Gallery, Bodleian Library, and Wellcome Trust Medical Photographic Library for permission to print illustrations and photographs.

Christine would like to thank her father, James Jespersen, for his thought-provoking conversation and his willingness to accompany her, on Father's Day no less, to the Body Worlds 2 exhibit. She would also like to thank her mother and stepfather, Lou Ann and Jon Shirley, for keeping a home where reading was a welcomed activity. In addition, she would like to acknowledge the many people who encouraged her in the project and helped her to make connections with contributors. She is indebted to Anne Becher, Dan Cress, Ann Dean, Katie Hogan, Calla Jacobson, Lisa Kay, Johnelle LaMarque, Vanessa Manhire, Ron Mallon, Bill Niemi, Rachel Stein, and Jim Westerman. She shows her appreciation to the CALL department for travel support and to faculty members who eased her workload. As always, thanks to Diane Dallin, Pamela and Daniel Farnsworth-Martin, Laura Haertling and Cody Andresen, Mia and Steve Mestdagh, Heather Orr, Deb Piranian, Gretchen and Raul Puente-Reinhardt, Meg Ryan, Dawn Skorczewski, and Julia Willis, whose friendships and hard questions have given her thought and sustained her throughout the years. Her sincerest appreciation goes to Alicita Rodríguez for conceptualizing and organizing the project. To both Alicita Rodríguez and Joseph Starr, she gives gratitude for their creative and clever thought, their exacting standards, and their abundant generosity and friendship. Special thanks to her husband, David Plante, for his love, intelligence, knowledge, scholarship, integrity, and ironic wit; without him, she could not have completed this book.

Alicita Rodríguez and Joseph Starr thank their parents, Alicia Martí Rodríguez, Ovidio Rodríguez, and Jacqueline Starr for their support. They also thank Michael Sappol of the History of Medicine division of the National Library of Medicine and Tony Walter of the Department of Social and Policy Sciences, University of Bath, for their scholarship on Body Worlds and their interest in and suggestions regarding our book. Thanks to Christine Jespersen for being a reliable, intelligent, and encouraging colleague; a faithful friend; and a conscientious and respectful editor; but especially for being the nicest person they know. Finally, Joseph thanks Russell Edson for permission to reprint "Skeleton."

Table of Contents

Section Three
The Resurrection of Excoriated Bodies

Section Four
Disharmonious Bodily Openings

Section Five
Aesthetic and Instructive but Not Morally Offensive

Introduction

Alicita Rodríguez and Joseph Starr

Then one day, I was in the butcher shop in the university town where I was studying, and as I watched the sales woman slice ham, it dawned on me that I ought to be using a meat slicer for cutting kidneys.

— Gunther von Hagens[1]

Gunther von Hagens, the inventor of plastination — the method of preserving corpses by injecting them with plastic — and the mastermind behind the widely successful Body Worlds exhibitions, seems to receive inspiration from a variety of disparate sources. The butcher shop catharsis evokes the more violent, Frankensteinian aspect of his anatomical craft, but von Hagens relies just as easily on more respectable justifications. For example, he emphasizes the pedagogical nature of his international traveling shows, claiming that he wants to "democratize anatomy." In fact, democracy, as part of his desire for political freedom, helped create the Body Worlds project. Von Hagens cites his two-year prison term in East Germany as a pivotal moment: he "made a clear decision in favor of uncertainty, because uncertainty always represents an opportunity"[2] — becoming an inventor being the opportunity, in this case.

While von Hagens' presentation of his patented methods of bodily preservation, both in his exhibitions and his writings, are often suspect, his scientific innovation and intricate craftsmanship cannot be denied. Nor can the marketing success of his enterprise. As of early 2008, since its opening in Japan in 1995, over twenty-five million visitors have attended the exhibitions in North America, Europe, and Asia.[3] Several derivative exhibits, which the Body Worlds website calls "copycat" shows, have sprung up in the wake of the venture's sensation,[4] and the exhibits have boosted attendance at American museums to record numbers. While the American Association of Museums collects no data on total attendance for specific exhibits and reports no such national or international database, individual museums confirm record attendance to von Hagens' shows. Most significantly, the California Science Center, which hosted the United States debut of Body Worlds 1 in July of 2004, reported that during its 205 day run, the exhibit attracted more than 665,000 visitors — the largest attendance for a non-permanent exhibit in the history of that museum.[5]

The continued popularity of successive incarnations testifies to the show's staying power: Body Worlds 2 opened at the same museum a week later, attracting an additional 265,000 visitors over a mere 57 days.[6] Figures at different locales are equally impressive. The Denver Museum of Nature and Science states that Body Worlds 2, which showed in Denver for 4.5 months, attracted 650,000 visitors, the third largest number in its history, with only Rameses II and the Aztec exhibits inviting larger crowds.[7] And at the Boston Museum of Science, the exhibit drew 550,000 visitors during its five month stay in late 2006.[8]

The consistent popularity of the traveling exhibit suggests that it is a cross-cultural enterprise. While the marketing strategy of Body Worlds may evolve with each new location, the central allure does not: people of all classes and ethnicities are drawn to von Hagens' plastinates, are compelled by his picturesque dead. As Stephen Colbert might say, the market has spoken in regard to Body Worlds: it is a bona fide world event.

The creator behind this endeavor is a marketing genius, albeit a Janus-headed figure, at once welcoming and shunning publicity. The Institute for Plastination (IfP), the corporation behind the Body Worlds exhibits, controls its image by censoring the media even as it feigns openness. Through the official Body Worlds website, one may download up to three high-resolution photographs; of the images available, however, none shows the more controversial whole-body plastinates (such as the *Reclining Pregnant Woman* or the fantastical *Mystical Plastinate*). Many attempts by the editors of this collection to contact Body Worlds' publicity agents were ignored, which unfortunately forbade us from including unauthorized photographs, and at least two of the science museums hosting Body Worlds in the United States kindly refused our requests for additional materials such as text of exhibition signage. If von Hagens can exercise his right to free speech, and, furthermore, if he can champion such as a core notion of his enterprise, why should he prevent others from doing likewise? Why should he constrain access to the tangible components of his mission to "democratize anatomy?"

Strangely enough, Gunther von Hagens has been known to court controversy, "escort[ing] guests dressed as space creatures through the exhibit with TV cameras in tow"[9] and performing a public autopsy in London in 2002.[10] His signature black fedora, which he seemingly never removes, also marks him as something of a showman. But von Hagens tempers every theatrical gesture with a didactic one. He is artist and scientist all at once, inventor and educator, individualist and philanthropist.

To this impressive resume of professions and identities, one is tempted to add linguist, for the man has a way with words. Though his aptitude for the pseudo-scientific turn of phrase, the anatomical coinage, the odd juxtaposition of terminology from medicine, the arts, and alchemy, might be attributed to his invention of a new science and a new art — a thing heretofore unseen — it seems owed in equal measure to his idiosyncratic personality. Some emblematic "von Hagenisms" include "forced impregnation," "unadulterated authenticity," and "a souvenir of mourning."[11] But we've saved our favorites for our section headings. To the caveat that this choice is an unfair characterization, that it is a re-contextualization that distorts meaning, we'd reply that von Hagens' curious utterances speak more truth than he may know, that the way he uses language exposes certain hypocrisies and contradictions, opening, in the process, assorted fissures in the monolith of his concerted attempt to control meaning.

Section One, "A Beauty Salon for Bodily Interiors,"[12] focuses on the ways in which the American incarnation of Body Worlds is constructed to elicit a constrained response. The phrase is taken from a passage in which von Hagens characterizes the typical anatomized cadaver ("partial dissections or individual specimens preserved in formaldehyde")[13] as less intrinsically instructive than his plastinates, which "will positively amaze viewers acquiring knowledge in such an aesthetic-instructive manner that they will find the design compelling just because it is convincingly aesthetic."[14] Each of the essays in this section investigates the rhetorical devices the Body Worlds' purveyors employ in order to manipulate viewer response. Joseph Starr's "The Plastinates' Narrative" looks at some of the verbal slippage in the Body Worlds' press material, logical sleights of hand that evade and obfuscate so as to convince consumers of the exhibit's validity as an apparatus of scientific instruction. Natalia Lizama's "Afterlife, but Not as We Know It: Melancholy, Post-Biological Ontology, and Plastinated Bodies" defines the status of the plastinate as "post-biological," asserting that plastination removes the corpse from the natural processes of decomposition or chemical destruction, thus imbuing it with a profoundly melancholic aspect that influences viewer response. In "Locating the Sublime," Lisa Nevárez reads the exhibit through the lens of Edmund Burke's theory of the Sublime, proposing that—contrary to von Hagens' intentions—exhibit-goers do not experience the plastinates as former persons or as altered forms of the once-human, but as representational art forms that elicit the somewhat ironic sensation of constrained boundlessness that characterizes the sublime. Aviva Briefel's "Take Me: The Rhetoric of Donation" is a close reading of the Donation Consent Form. Comparing the language of the form with some narrative devices of gothic horror, Briefel asserts that the document engages in a particularly cunning form of linguistic trickery, one in which the prospective body donor is promised a sort of posthumous jurisdiction over her plastinated corpse. Lastly, Rebecca Onion's "Individualist Etiologies: Environmental Health, Biological Risk, and Medical Display" exposes how the predominately clinical tone of the exhibit breaks down notably in the smoker's lung and obesity displays, at which point the language becomes didactic and exhortative, sanctimoniously admonishing "lifestyle choices" that lead to poor health, while ignoring the pervasive environmental degradations that make the promised sovereignty over individual health a patent falsehood.

Section Two explores the antecedents of Body Worlds, the preserved cadavers and specimens that von Hagens refers to as "the usual gruesome anatomical apparitions."[15] In "The Persistence of Tradition in Anatomical Museums," Stephen Johnson traces Body Worlds' lineage to Victorian anatomical museums and examines why von Hagens simultaneously echoes and rejects those associations. Johnson also compares the European and American incarnations of the exhibits, which playfully parody and somberly reject their historical predecessors. Pedro Ponce's essay, "Illuminating the Soul: Panopticism and the Freak Show," takes the idea of ancestry one step further, theorizing that Body Worlds continues the freak show tradition, exploiting our voyeuristic desire for the carnivalesque. By juxtaposing von Hagens' plastinates and Bentham's Auto-Icon, Patricia Pierson considers the role of collecting and the narrative that specific collections construct. "The Amethyst Seal: Anatomy and Identity in Bentham and von Hagens" ultimately poses that Body Worlds is a form of thanatology.

Section Three, "The Resurrection of Excoriated Bodies,"[16] concerns Body Worlds' engagement with European historical contexts including the motif of bodily degradation and the rise of abstractionist art. Natalie Loveless' "Affecting Bodies" interprets Body Worlds as counter-propositional to Paul Virilio's notions of "monstration," in which abstraction is decried, claiming instead that the art object inheres not in the plastinate itself, but in the moment of affective encounter between plastinate and witness, a confluence that is rife with pedagogical potential. Peter M. McIsaac's "Worrying About Democratic Values: Body Worlds in German Context, 1996–2004" examines how von Hagens' narrative of personal integrity and authenticity is in fact a coercive device intended to obfuscate the ways the exhibit resonates with Nazi and Holocaust remembrance, a blocking apparatus in parallel with certain features of contemporary German democracy. Likewise, Alexandra Ludewig in "The Echo of German Horror Films" engages the pervasive hold that the Holocaust continues to have on the German imagination, presenting the horror film *Anatomie* as an allegory of an ethos of medical experimentation —fathered by the Nazis— whose contemporary incarnation is bound up with von Hagens and a "scientific knowledge at any cost" kind of aesthetic. The final essay in this section, Uli Linke's "Touching the Corpse: The Unmaking of Memory in the Body Museum," analyzes the global commodification of memory, pain, and suffering, characterizing the modern museum as the most salient repository of these bodily artifacts, and the German roots of von Hagens and Body Worlds as a particulary charged manifestation of this new commerce.

Linke's discussion of the gendered aspects of Body Worlds also serves as an introduction to Section Four, "Disharmonious Bodily Openings,"[17] a category of perceptions of the human body that von Hagens lists under the column "Ugliness."[18] In truth, von Hagens dislikes the messier characteristics of physiology such as fluids and odors— natural symptoms of mortality in general and perhaps (in his mind) of femininity in particular. T. Christine Jespersen and Alicita Rodríguez's "Forced Impregnation and Masculinist Utopia" deconstructs the gendered universe created by von Hagens with his gestalt plastinates, focusing on a close reading of the staging of Body Worlds 2, including the fetal development room. The role of the fetus is further investigated by Christian DuComb in "The Politics of Fetal Display," which decodes the cultural meaning of fetuses. Elizabeth Simon Ruchti closes this segment with "Corpse-less: A Battle with Abjection," an analysis of the plastinates' queer and abject properties, both those identified and shunned.

"Aesthetic and Instructive but Not Morally Offensive"[19] is the book's last section, a look at the ethical implications of plastination. The title of Section Five comes from von Hagens' explanation of the beauty of his plastinates: he thinks that their aesthetic appeal eschews moral objections, which result from visceral reactions to the grotesqueness of traditional cadavers.[20] But ugliness has little to do with the bioethical concerns raised by the concluding essays. Ruth Levy Guyer concentrates on the provenance and consent of the donated bodies in "Anatomy Without Integrity," a subject that serves as a starting point for Paul J. Wojda's "Adoration, Veneration, Plastination: Theo-Liturgical Reflections," which ultimately analyzes the spiritual meaning of the dead body from a Catholic perspective: what can ethically be done to the mortal shell that houses the soul? The essay, "Twilife: The Art and Science of Consuming Death," written by Lucia Tanassi, examines how von Hagens creates a new sociology of death, one that involves a commodification of biology. Finally, Ara Osterweil and David Baumflek's "Emergent Bodies: Human, All Too Human, Posthuman" situates Body Worlds within a larger context of body art, arguing that it radically depoliticizes the dead.

Notes

1. von Hagens, "Anatomy and Plastination," 26.
2. Ibid.
3. Institute for Plastination, "Prelude."
4. Institute for Plastination, "Original and Copycat."
5. California Science Center, "Gunther von Hagens' Body Worlds 2."
6. Ibid.
7. Aguilar, "Body Worlds 2"; Denver Museum of Nature and Science, e-mail.
8. Noonan, "*Bodies*— Educational or Exploitative?"
9. Moore and Brown, "Gunther von Hagens and Body Worlds: Part 1: The Anatomist as Prosektor and Proplastiker," 211.
10. Ibid.
11. von Hagens, "Anatomy and Plastination," 21; *Ibid.*, 33; "On Gruesome Corpses, Gestalt Plastinates and Mandatory Interment," 279.
12. von Hagens, "On Gruesome Corpses," 268.
13. Ibid.
14. Ibid.
15. Ibid.
16. Ibid., 272.

17. Ibid.
18. Ibid., "Tab. 1: *Perceptions of the human body*."
19. Ibid., 268.
20. Ibid., 267–8.

Bibliography

Aguilar, John. "Body Worlds 2 Packed Up for Boston." *Rocky Mountain News*, July 24, 2006. http://www.rockymountainnews.com/news/2006/jul/24/body-worlds-2-packed-up-for-boston/ (accessed March 15, 2008).

California Science Center. Press Release. "Gunther von Hagens' Body Worlds 2: The Anatomical Exhibition of Real Human Bodies." Los Angeles, 2005. http://www.californiasciencecenter.org/GenInfo/MediaRoom/PressReleases/BodyWorlds/BW2.php (accessed March 18, 2008).

Denver Museum of Nature and Science, e-mail message to Patrick Muckleroy, March 8, 2008.

Institute for Plastination. *Donating Your Body for Plastination*. Seventh Edition. Heidelberg: 2004.

_____. "Original and Copycat." Gunther von Hagens Body Worlds: The Original Exhibition of Real Human Bodies. Heidelberg: 2006. http://www.bodyworlds.com/en/exhibitions/original_copycat.html (accessed March 15, 2008).

_____. "Prelude." Gunther von Hagens' Body Worlds: The Original Exhibition of Real Human Bodies. Heidelberg: 2006. http://www.bodyworlds.com/en/prelude.html (accessed March 15, 2008).

Moore, Charlene M., and C. Mackenzie Brown. "Part I: The Anatomist as Prosektor and Proplastiker." In Whalley, *Gunther von Hagens'*, 208–13.

Noonan, Erica. "*Bodies*—Educational or Exploitative?" Boston Globe, November 8, 2007. http://www.boston.com/news/local/articles/2007/11/08/bodies___educational_or_exploitive/?page=2 (accessed March 18, 2008).

von Hagens, Gunther. "Anatomy and Plastination." In Whalley, *Gunther von Hagens'*, 9–37.

_____. "On Gruesome Corpses, Gestalt Plastinates and Mandatory Interment." In Whalley, *Gunther von Hagens'*, 260–82.

Whalley, Angelina, ed. *Gunther von Hagens' Body Worlds: The Anatomical Exhibition of Real Human Bodies*. Translated by Francis Kelly. Heidelberg: Arts and Sciences, 2006.

Section One

A Beauty Salon for Bodily Interiors

The Plastinates' Narrative

Joseph Starr

Gunther von Hagens' Body Worlds supposes a new way to *look* at our bodies, in both the literal and figurative senses of the verb, since the rhetoric of the exhibition's custodians, and, notably, of von Hagens himself, contends that plastinated corpses (or "plastinates," for von Hagens, who would prefer to give his anatomical models their own identities, distinct from the loaded significations of "corpse" or "cadaver") "democratize anatomy":

> Body Worlds allows people from all backgrounds to better understand the body and how it functions ... as event anatomy, Body Worlds juxtaposes healthy and diseased organs and features whole body specimens in dynamic lifelike poses. Plastinated anatomy is beauty beneath the skin, frozen in time between death and decay. It opens the heart to our inner self and makes us fall in love with our own body.[1]

In von Hagens' view, this new self love is distinct from today's cult of the body—the exercise-sculpted physiques and surgically altered faces that dominate contemporary media culture. Nor does it resemble the self-help paradigm with its promise of spiritual affirmation and subsequent boost to self-esteem. Von Hagens' "inner beauty" follows a reverse course: to witness the purely aesthetic appeal (and in Body Worlds this is an aesthetic that can only be "seen," never touched, tasted, or smelled) of muscles and bones, organs and nerves is to adopt a broadly humanistic outlook, in which apprehending the universal beauty of our physical inner self allows us all to be more comfortable in our own skins, and by extension, more empathetic—better-informed, kinder human beings.

It's all very progressive, utopian even, which isn't surprising, since von Hagens tends to align himself with both Renaissance anatomists and Enlightenment thinkers in the sense that they wished to deconstruct and demystify the body, thereby easing the layperson's apprehensions surrounding his corporeality, apprehensions that point toward a fear of life as well as of death.

In response to contentions that Body Worlds is in some degree exploitative, von Hagens employs a dualistic rhetoric that aligns his critics with the religiosity of the dark ages, a contingent that is manifest today as "the defenders of Terri Schiavo, the opponents of euthanasia, and the dissenters from assisted suicide."[2] In an opinion piece printed in the *London Times Higher Education*

Supplement, von Hagens claims that his exhibits "answer a great need in modern society to understand our bodies":

> I think it is high time for Britain to democratise its bodies and the functions of the body. The denial of death is a modern phenomenon of the medical industry. Until the end of the 19th century there were anatomical displays in public theatres all over Europe — but now ordinary people are not allowed to see them. I think my exhibition, which people pay an entrance fee to see, is more democratic than anatomical education.... Hollywood has done a lot to associate anatomy with the horror and emotional distress of dying. But my work is very different. I put specimens in a celebratory setting. The aesthetic helps people to understand the body. I see it as an enlightenment, which is a part of science. Far from degrading anatomy, I think my work dignifies it. It is much more dignifying than the conventional use of specimens, in which body parts are cut up and put in jars. The unique process of plastination, in which the liquids and tissue of the human body are replaced with polymers giving them a fresh, plastic appearance and texture, makes them beautiful.[3]

It's curious that von Hagens evokes Hollywood as an oppositional construct, not only because his talk of "enlightenment" and "democratism" resembles the kind of easy/idealized liberalism most commonly associated in the public imagination with a celebrity intelligentsia, but also because the Body World's website features approbatory comments from the rich and famous. To wit,

> An absolutely fascinating experience! Only this morning I was discussing with my partner how unique we are as human beings and this has proved it too me. Why fight wars? What is the point of racism? We are all the same! Thanks!— **David Harewood**, UK actor, London 2002
>
> It is amazing and enlightening. A fascinating learning experience. I will never take my body for granted again.—**Tony Hawk**, U.S. Skateboarder, Los Angeles 2004
>
> Thank you! I am very grateful to have been invited to this exhibition. It was the most gorgeous meeting of the scientific and the poetic. How humbly the maker bowed to science made the greatness of this so much larger. *Gratitude* for the nourishment and the generosity.—**Björk**, Icelandic singer, Brussels 2001[4]

These de-facto cultural authorities thus affirm some of von Hagens' key contentions: 1. Body Worlds eliminates the perception of difference; 2. Body Worlds prompts us to take better care of our bodies; 3. Body Worlds harmonizes art and science. If the celebrities' claims are facile, they are no less so than von Hagens' defense of himself and his exhibits, utterances that begin to resemble the sanitized press releases of a politician, the institutionalized party-line responses— all geared toward "damage control"— of a figure-head protecting an oft-maligned enterprise. With Body Worlds entering its eleventh year, von Hagens has become an old hand at defending himself. Since 1997, he's had to answer to charges that, among other offenses, his exhibits desecrate the memory of the dead (or important cultural beliefs about the sanctity of death); ignore or minimize issues surrounding consent for donation (it is alleged that some of his plastinates were Chinese nationals executed for political crimes);

feature gratuitous manipulations that aim to shock rather than educate; and revive the historical conflation of medicine with murder, classism, racism, and sexism.

Von Hagens has a patented response to each criticism: the severing of the paper trail that could link person to plastinate is necessary to preserve the anonymity so crucial to maintaining scientific objectivism; the playful poses of some of the plastinates calm the spectators by bringing "death closer to life in a kind of humorous fashion"[5]; the dearth of female whole-body plastinates is owed to females' inferior musculature.

Most crucially, he habitually cites Body Worlds' widespread acclaim in terms of high-attendance figures and ticket receipts, polled responses of attendees, and the growing numbers of volunteers who wish to donate their bodies for plastination. In a discussion on his affinities with the Renaissance anatomist Josef Beuys, von Hagens refers, somewhat ironically, to this approbation of the masses: "the difference between us lies in the fact that he sought the favor of laypersons but won the approval of intellectuals, while I have courted the favor of professional colleagues but have actually only gained the approval of laypersons."[6]

Owing to the enormous revenue-generating machine that the plastinates and Body Worlds have created, von Hagens is (or has become) a consummate showman. Though he admits as much, "what's wrong with sensationalism? In today's media-saturated world, anatomy needs a little shock and awe, and some humor, just to get people to walk through the door and pay attention,"[7] his insistence on the didactic/educational value of Body Worlds seems somewhat disingenuous. Von Hagens' defense of his exhibitions is largely predicated on a supposition of scientific objectivity — i.e., that his displays of plastinates facilitate an unmediated "universal access" to "real" human bodies; that his techniques preserve an objective reality (the reality of the just-dead, but not yet decomposing body); that he has taken pains to remove the corrupting influence of emotion that might be evoked by the suggestion of the human identity/personality of his plastinates. In short, that he merely fulfills a need for which the inhabitants of a free society clamor. The problem with this formulation is that it does not recognize — or, worse, attempts to efface — the extraordinary degree to which Body Worlds is a constructed enterprise. This is an instance of the "objectivity fallacy" that frequently discredits the supposed impartiality of scientific experimentation: for nearly every aspect of Body Worlds is geared towards eliciting a constrained response — that of admiration for the pleasing aesthetic qualities of von Hagens' plastinates.

Whether Body Worlds' collection of organs, bones, bodily cross-sections, body fragments, and whole-body plastinates is pleasing to look at is largely a subjective determination. Most people seem to agree that the whole-body plastinates, which are invariably the stars of the show, have a pleasant aesthetic; likewise, the displays of the nervous and respiratory systems— which, in Body Worlds 2 are presented against a dark background and suffused with a fluorescent dye—conform to some notion of universal beauty. But it's difficult to prettify an organ (even a healthy one), and in spite of press material that alleges that "plastinated anatomy is beauty beneath the skin,"[8] I doubt the veracity of anyone who claims to enjoy looking at the human digestive tract—plastinated or not.

Indeed, there seems to be some kind of mass denial at play here, emanating from the exhibition's creators and adopted by its spectators, that everything inside us (once plastinated) is consummately beautiful. This is the inscribed narrative created by von Hagens and his purveyors, propagated by the Body Worlds' website and press material, and embraced by the spectators who rapidly learn that the *appropriate* response to the exhibition—in spite of von Hagens' claims to playfulness—is a sort of stunned and amazed reverence.

Of course, as all architects of public consensus are aware, it would be impossible to achieve such consensus if certain of the contentions didn't ring true. This is no different with Body Worlds, and in von Hagens' defense, there's scant disputing that he's a talented scientist, anatomist, tactician, and artisan. Several medical authorities who have spoken publicly about Body Worlds affirm that his dissections are expertly performed (see Fitzpatrick, "Making a Spectacle of Ourselves," and Stolkin, "The Anatomists"). Further, the scientific value of plastination is widely affirmed: as von Hagens contends, the technique facilitates an improved understanding of the spatial relationships of human anatomy; it allows us to create anatomical models that can be studied without being degraded, that can be manipulated in a way that allows a very specific vantage, that are durable and somewhat permanent and can be transported without being destroyed. It is this last quality that makes Body Worlds possible, since whole-body plastinates are hardly disposable, each one representing countless hours of labor and substantial expense.

As the Body Worlds' literature makes clear, the process of plastination involves considerable alteration to the corpse or "recently deceased" or "preplastinate" if you prefer: "Plastination ... replaces bodily fluids and fat with reactive polymers such as silicone rubber, epoxy resins or polyester ... plastinated specimens are dry and odorless; they retain their natural surface and are identical with their state prior to preservation down to the microscopic level."[9] This is one way to characterize plastination; another is to say that the cadaver's water content (which in a typical specimen constitutes 70 percent of the body mass) is ultimately replaced with plastic. In other words, 70 percent of the original material becomes something else entirely. By any reasonable standards of appraisal, once plastinated, the body is no longer a body: it is, rather, an identical rendering, a faithful three-dimensional map, a life-sized anatomical model, but no longer a human body. The majority of the physical stuff—the messy, smelly, gooey liquidity that is endemic to life—is diffused into the atmosphere during the initial cold solvent bath.

One difficulty with von Hagens' narrative is that he seems to want it both ways. That is, while he claims that Body Worlds offers the allure of "authentic" or "real" human bodies, he yet denies this humanity in his endorsements of plastinates as "dry," "odorless," "plastic," "shiny," and "clean."[10] The objection aims to call attention to a hypocrisy in the Body Worlds' narrative; namely, that the "bodies" are both human and non-human — a patent impossibility. But to admit that plastinates are something else entirely — that they are, in fact, not human but rather an ingenious anatomical innovation that *converts* bodies into three-dimensional models— would be to defeat the narrative that culminates in awe and reverence for the presentation of "real human bodies."

A second fallacy in the Body Worlds' narrative is that the plastinates facilitate a comprehensive experience of human anatomy. Nearly every aspect of the Body Worlds exhibitions is drawn in visual terms: the hyper-real appearance of plastinated muscle (which is featured more prominently than bones, organs, connective tissue, etc.); the medical "lessons" on healthful living, which feature displays comparing diseased organs with healthy ones; the museum setting, with its intimations of civilized decorum, of sacred spaces that house valuable and fragile objects of art. In a paper entitled "Animated Corpses: Communicating with Post Mortals in an Anatomical Exhibition," Stefan Hirschauer appraises the ways in which Body Worlds is designed to appeal to a single sense: "Like all anatomical representations, the Body Worlds (sic) works with a suggestion of *visibility* that suppresses all dimensions of bodily experience which cannot be mediated through the eye.... Dryness and lack of odour guarantee that this would remain a detached visual experience, allowing a safe curiosity about the body."[11] The spectators' curiosity is "safe" because Body Worlds homogenizes death — a homogenization that emerges in spite of the varied poses, the "playful" accoutrements, even the dramatic "explosions" or "lateral expansions" that showcase how the body's constituents relate to one another spatially — in the sense that it creates a uniform visual experience and in the sense that it creates a ritualized, communal happening in which the morbidity typically associated with death is suppressed. As Hirschauer rightly suggests, von Hagens achieves this suppression by characterizing the experience of the inner body as an exclusively visual event.

The argument that von Hagens' plastinates are superior anatomical specimens, though true in some regards, is vulnerable here as well, since their utility to the anatomist or student of medicine is also limited. Dissection, though it may be messy and even "undignified," invites a true bodily experience of death; that is, the student experiences the cadaver as a thing to be touched, heard, smelled. Certainly the ability to remove an organ from its bodily niche, to feel the weight of it, to see and hear how it inhabits its cavity and how it is positioned and fixed in relation to the surrounding constituents is intrinsically useful, is a truer approximation of what happens inside living bodies.

Lastly, the skinless-ness of von Hagens' plastinates invites critique. Though this feature is in some sense a pragmatic necessity, since Body Worlds supposes the experience of seeing *inside* the body, were Body Worlds a truer representation of anatomy, skin would be a more prominent feature. In "Getting Under the Skin, or, How Faces Have Become Obsolete," Bernadette Wegenstein remarks on the literal and figurative import of the body's largest organ:

> The skin has the most weight (18 percent of an adult's weight), and occupies the largest surface (18,000 cm^2) of all our organs ... [it] serves as an "interface" between "me" and the "other" by, on the one hand, protecting the ego like an envelope, and on the other hand, dividing it from the outer world ... [it] is the organ that needs to be penetrated at first in order to get into the body; it is the organ that sur-faces the body, and just as in a synechdotal trope, it constitutes the entrance to the inner body universe: the skin is the border between the outer and the inner body.[12]

It is also the organ responsible for touch, which is arguably our most substantive means of engaging with the world.

The only whole body plastinate in Body Worlds that acknowledges the centrality of the skin is *The Skin Man*, a specimen which Michael Lewis characterizes as a "doleful cadaver hold[ing] up his own flayed skin as if deciding whether to send it to the cleaners."[13] Contrary to Lewis, I don't read *The Skin Man* as "doleful," but rather as contemplative, even enlightened, for having divested himself of this unruly garment — a reading in keeping with von Hagens' conception of "beauty beneath the skin." And this plastinate does appear vital: the rich red-pink of his pronounced musculature is a vivid contrast to the pasty, crumpled skin suit he seems poised to cast away.

In an overt way, *The Skin Man* announces Body Worlds' intention to explore the body's interiors, yet it also reveals von Hagens' prejudice toward the visually pleasant. In an exhibition which purports a comprehensive exploration of human anatomy and a concomitant dismantling of taboos about bodily interiority, one might expect the skin to occupy a central position: for what other body part better evokes the metaphor of communion between self and other? Of transgressing social as well as private mores? Of the intimacies of birth, sex, and medical intervention? In the same way that some of the plastinates feature "expansions" that allow simultaneous viewing of the muscular and digestive systems, for example, von Hagens could incorporate skin as a prominent motif without compromising the principal aim of interior investigation.

But plastinated skin is ugly. Without the supportive and lubricative functions of bodily fluids, human skin looks pale, formless, wrinkly — dead. In both the literal sense of the appearance of a corpse prepared for dissection, as well as the figurative sense of the gruesomeness and morbidity commonly associated with the physical aspect of death, plastinated skin is too evocative of von Hagens' bugaboo, the cadaver. One only need view the collection of wholly-skinned fetuses displayed in the "pregnancy room" to confirm this assertion.

The prose poet Russell Edson has a piece that is relevant to this discussion:

> A skeleton awakens and yawns in a grinning gape of teeth; stretching, its spine arching like a suspension bridge.
>
> Sitting on a coffin it pulls flesh as if hip boots up the bones of its legs. Toe bones wriggling and twisting to fit themselves into the flesh of their feet.
>
> A long sleeved glove of flesh it pulls up one arm; and then the other.
>
> And so with its torso, wrapping its ribs in a corset of flesh, fastened with the stub of an old umbilical cord.
>
> Almost complete it hangs its genitals and puts on a mask.
>
> In the street it passes unnoticed by other skeletons also wearing masks.[14]

I quote "Skeleton" because Edson's playful if somewhat macabre treatment of the ways in which our bodies both protect us and constrain us is sincere in a way that Body Worlds is not: the piece is an apt characterization of what a Body Worlds' experience could be, how it might foster an honest and productive appraisal of the problematic relationship we all have with our bodies, with the bodies of others, and with death — were it not for the imposition of von Hagens' narrative of reverence and awe.

Notes

1. "Discover What You're Made of When Gunther von Hagens' Body Worlds 2: The Anatomical Exhibition of Real Human Bodies Opens at the Denver Museum of Nature and Science."
2. Lewis, "Body and Soul," 5.
3. von Hagens, "Why I think it is high time we democratised the science of anatomy."
4. Institute for Plastination, "Celebrity Comments."
5. von Hagens, "On Gruesome Corpses, Gestalt Plastinates and Mandatory Interment," 262.
6. Ibid., 273.
7. Bohannon, Ding Yimin, and Xiong Lei, "Anatomy's Full Monty," 4.
8. "Discover What You're Made of."
9. Ibid.
10. "Discover What You're Made of," passim; "On Gruesome Corpses," passim.
11. Hirschauer, "Animated Corpses: Communicating with Post Mortals in an Anatomical Exhibition," 32.
12. Wegenstein, "Getting Under the Skin, or, How Faces Have Become Obsolete," 251, 239.
13. Lewis, 5.
14. Edson, "Skeleton," 49 (reprinted in its entirety with permission of Russell Edson and *Sentence.*)

Bibliography

Bohannon, John, Ding Yimin, and Xiong Lei. "Anatomy's Full Monty." *Science* 301, no. 5637 (2003): 1172–1175.

"Discover What You're Made of When Gunther von Hagens' Body Worlds 2: The Anatomical Exhibition of Real Human Bodies Opens at the Denver Museum of Nature and Science." Body Worlds (Press Release). Heidelberg: Institute for Plastination, December 14, 2005.

Edson, Russell. "Skeleton." *Sentence* 4 (2006): 49.

Fitzpatrick, Michael. "Making a Spectacle of Ourselves." Spiked Online. Health. London: Signet House, March 26, 2002. http//www.spiked-online.com/Articles/00000006D862.htm (accessed February 10, 2007).

Hirschauer, Stefan. "Animated Corpses: Communicating with Post Mortals in an Anatomical Exhibition." *Body and Society* 12, no. 4 (2006): 25–52.

Institute for Plastination. "Celebrity Comments." Gunther von Hagens' Body Worlds: The Original Exhibition of Real Human Bodies. Heidelberg: 2006. http://www.bodyworlds.com/en/exhibitions/celebrity_comments.html (accessed February 10, 2007).

Lewis, Michael J. "Body and Soul." *Commentary* 123, no. 1 (Jan 2007): 29–32.

Stolkin, Colin. "The Anatomists." Channel 4. London: 4 Digital Group. http//www.channel4.com/Science/microsites/A/anatomists/opinions5.html (accessed February 10, 2007).

von Hagens, Gunther. "On Gruesome Corpses, Gestalt Plastinates and Mandatory Interment," In Whalley, *Gunther von Hagens'*, 260–82.

_____. "Why I Think It Is High Time We Democratised the Science of Anatomy." *Times Higher Education Supplement,* March 22, 2002. http://www.timeshighereducation.co.uk/story.asp?storyCode=167963&sectioncode=26 (accessed February 10, 2007).

Whalley, Angelina, ed. *Gunther von Hagens' Body Worlds: The Anatomical Exhibition of Real Human Bodies.* Translated by Francis Kelly. Heidelberg: Arts and Sciences, 2006.

Wegenstein, Bernadette. "Getting Under the Skin, or, How Faces Have Become Obsolete." *Configurations* 10, no. 221 (2002): 221–259.

Afterlife, but Not as We Know It: Melancholy, Post-Biological Ontology, and Plastinated Bodies

Natalia Lizama

> *What we call the mortal remains evades the usual categories: something is there before us that is neither the living person himself nor any sort of reality.*
>
> — Maurice Blanchot, "Two Versions of the Imaginary"

One of the more compelling figures from Gunther von Hagens' Body Worlds exhibitions is *The Skin Man*, a whole-body plastinate who holds aloft his entire disembodied skin. His facial expression and gesture suggest a sense of bewilderment and awe, as though he is taken aback to discover that he has been flayed and the skin he holds is in fact his own. While the majority of plastinates in the exhibition are presented in poses depicting athletic activities, *The Skin Man* is one of a number of Body Worlds plastinates whose "activity" is one of philosophical self-reflection. Like *The Skin Man*, *The Thinker* and the human figure in the *Rearing Horse with Rider* (henceforth referred to as *The Rider*) also gaze upon fragments of human bodies, seemingly ruminating upon the nature of human being; *The Thinker* gazes upon a human skull while *The Rider* scrutinizes his own brain. These gestures of simulated human consciousness suggest that the plastinates are marveling at the intricacies of flesh and the wonder of human life. Yet these plastinates are imbued with a poignant melancholy that pertains to their liminal ontological status and their simulated consciousness. This form of melancholic longing can be seen in all of the Body Worlds plastinates but is heightened in the case of plastinates whose poses suggest self-awareness. Where melancholy is the irresolvable longing for an irretrievably lost object or ideal, the object of melancholic loss for the plastinates is organic death and decomposition, a process prohibited by plastination. Consequently, the Body Worlds plastinates are suspended in a kind of ontological limbo. Although derived from corpses, they will neither decay nor decompose; they are immobile like a corpse yet the quietude of death is disrupted by their animated poses that suggest vitality and health. In the context of their simulated consciousnesses, plastinates such as *The Skin Man*, *The Thinker*, and *The Rider* evoke an imagined melancholy that results from the loss of death and corporeal dissolution.

Plastination strips the body of fat and fluid before saturating it with a polymer solution. This process renders the plastinated corpse rigid, "realistically" colored, and more importantly, impervious to decay; as von Hagens reassures readers of the catalog, visitors will not be "bothered by any kind offensive odours" emanating from the plastinates.[1] The Body Worlds plastinates are displayed skinned and partially dissected, often with muscles prized open to reveal internal musculature. Most of the plastinates are presented in lifelike poses, mimicking those of athletic activities such as running, skiing, or skateboarding, often with props such as a basketball or skis that accord with the posture. The poses of the plastinates are extraordinarily athletic, suggestive of living bodies at the peak of their physical prowess that have been momentarily frozen in action in a kind of skinless *tableau vivant*. The display of corpses in Body Worlds has been the subject of much controversy and has incited numerous and widely differing responses in exhibition visitors. Body Worlds has been praised by some for its allegedly inspiring presentation of the human body that is "a joy to see" and condemned by others as a "voyeuristic event [that] violates human dignity."[2] Certainly the experience of viewing the plastinated bodies would seem to elicit strong reactions in visitors—the exhibition catalog claims that 75 percent of surveyed visitors intended "to devote a significant amount of time to emotionally dealing with their inner thoughts and feelings after experiencing the exhibition."[3] The intensity of visitors' reactions to the exhibition is partly contextualized by examining the extraordinary presentation of the corpse in Body Worlds. For the layperson, viewing a corpse usually takes place in a limited number of settings, such as in funeral parlors or hospitals, where the viewer's response is framed by cultural expectations and rituals that typically command a sorrowful, subdued reverence towards the corpse. For most Body Worlds visitors, viewing a corpse is a novel experience, even more so when the corpse is not only plastinated but also dissected and presented in an attitude of vitality.

The presentation of the corpse in Body Worlds recalls Philippe Ariès' discussion of the gradual abandonment of mourning rituals in western culture throughout the nineteenth and twentieth centuries.[4] According to Ariès, the cessation of mourning rituals has resulted in a silencing of death such that the significance of death has gradually been annulled.[5] Death has become a clandestine event, a subject to be avoided rather than acknowledged through mourning rituals.[6] This obfuscation of death is evident in the display of the corpse at funerals: with its complexion artificially brightened by makeup, and its body concealed by clothes, the funereal corpse in its attitude of repose may well be imagined to be merely sleeping rather than dead. Thus, the funereal corpse is re-invested with a sense of life, as if a residue of vitality still remains within the body until it decays. This imbuement of the funereal corpse with life recalls Maurice Blanchot's theorization of the troubling ontological status of the corpse. Blanchot's theorization of the corpse is, like Ariès', somewhat limited by its Eurocentric perspective. As Blanchot notes, the corpse "evades the usual categories: something is there before us that is neither the living person himself nor any sort of reality."[7] As the incorporation of a once-living human being, the corpse is thus a source of profound anxiety as it threatens the very status of selfhood by highlighting its impermanence. In light of Blanchot's notion, the funereal conventions dictating that the corpse be presented as if in a transitionary state between slumber and decay serve to alleviate cultural anxieties about its ambiguous ontological status.

The presentation in Body Worlds of the corpse as lifelike takes this obfuscation of death to an extreme degree. The investment of the corpse with vitality that takes place during funereal rituals is amplified in Body Worlds, where the presentation of the plastinates in animated poses indicates a similar attempt to conceal their innate morbidity. Not only are the plastinated corpses fixed in poses suggestive of life, but they also embody an air of exuberance and health. Indeed, it is quite feasible that many of the plastinates in their posthumous sporting endeavors are far more animate in death than they ever were in life. Thus, one of the most extraordinary aspects of the exhibition is the visitors' proximity to human corpses and the simultaneous obfuscation of death. This is intensified by the way that plastination renders the excessiveness and abjection of death invisible. Although the exhibition visitors view human corpses in close detail, they are subjected to none of the other sensory experiences, often far less spectacular and far more unpleasant, that are usually concomitant with viewing the internal anatomy of a non-plastinated corpse; there is no smell of formalin, no wetness of blood or graying flesh, and, because the bodies are already dissected, there is no sound of flesh being prized open. Plastination renders the body anatomically spectacular whilst eradicating the abject qualities of the corpse and in this way transforms the body into a three-dimensional anatomical illustration; the plastinate is both spectacular and anatomically informative, yet is also clean, dry, and legible as if the body were neither dead nor dissected but merely rendered visible.[8]

This phenomenon is indicative of the way in which the exhibition, by re-investing the corpse with life, complicates the dichotomy between life and death. The lifelike poses of the plastinates suggest that the most important change in their transformation from human to plastinate is not death, but in fact plastination. That is, because of the plastinates' lifelike poses, viewers are positioned to imagine that they are being afforded a glimpse into the interior of a living body, rather than gazing upon a technologically modified corpse. Through the exhibition's focus on the plastinates' posthumous existence, death is thus rendered insignificant, and the modificatory aspect of plastination is concealed by the spectacular mode of visualization it affords. The plastinated bodies are presented as if plastination enables the anatomy of living human beings to be faithfully represented, rather than completely modifying the body's chemical and physical structure.

While the plastinates' dual embodiment of life and death serves in one sense to assuage cultural anxieties about the ontological status of the corpse, it also results in an incongruity between the plastinates' embodiment of differing states that would otherwise be mutually exclusive; namely, life and death. Through plastination the body is endowed with a form of synthetic immortality; it will not decompose or disintegrate but will retain its vivid, animated appearance and pose. Through plastination the body is corporeally resurrected and suspended indefinitely in a state of posthumous being somewhere between organic death and synthetic life.

This technologically-derived posthumous life of the plastinate recalls Hans Moravec's notion of "post-biological" ontology.[9] In his proposed transmigration scenario, robotics researcher Moravec envisions a future in which human beings have evolved to become completely compatible with technology.[10] Within this "post-biological" future, it will one day be possible to "transmigrate"; to download a human being's complete consciousness into a computer whilst, presumably, the remaining body is discarded.[11] This technologized realization of Cartesian dualism will, according to Moravec, enable human beings to be resurrected as fleshless entities existing entirely within the machine, thereby completely disembodying the "self."[12] The process of plastination, in which bodies are posthumously revived as synthetic, non-biodegradable beings, retains the dream of technologically assisted resurrection implicit in Moravec's fantasy. However, where Moravec imagines the body to be discarded whilst consciousness is cybernetically reconfigured, the process of plastination reverses this paradigm, reconfiguring the flesh whilst selfhood is all but eradicated. Through the removal of the skin, plastination renders the corpse unidentifiable; in its posthumous existence as plastinate, the body bears no resemblance to the person it once was, and the original "self" is disappeared into anonymity.

The plastinate thus exemplifies the post-biological in a very literal sense; it is a human body that, through the assistance of technological modification, exists beyond the parameters of biological human life. Through plastination, the body is resurrected as an anatomical artifact that retains no residue of actual personhood but is reinvested with a form of simulated subjectivity suggested by the body's athletic and self-reflective posture. The application of Moravec's term "post-biological" to the plastinates is thus intentionally ironic. The plastinates literalize post-biological ontology, while simultaneously subverting

Moravec's imagined form of immortality: post-biological existence for the plastinate is afterlife but not as Moravec would know it. It is through this subversion of the trope of resurrection, in which plastination revives the corporeal body rather than consciousness, that the melancholy of the plastinate derives. For the plastinate, the lost and irresolvable object of melancholy is the body as a living organism, one whose death results in decay and ultimate dissolution. Because of their simulated subjectivity, the plastinates are endowed with a corporeal immortality that retains none of the imaginary disembodied bliss of the afterlife: the lost ideal of melancholy is thus the promise of organic decomposition that death entails. In this sense, the sadness of the plastinate is that of the immortal: the peace of death will never arrive. Locked in a synthetic afterlife, the plastinate is condemned to endlessly re-enact its athletic gesture for a Sisyphean eternity.

Melancholy is defined in psychoanalytic discourse as the affect that results from an irresolvable sense of loss.[13] As Andrew Gibson in his study of melancholy describes the condition, "the melancholic is a mourner who cannot work his or her way through to the end of grief ... there is no resolution to, no purgation of ... no materialization of grief."[14] The sadness of melancholy is thus a nebulous one; the source of grief is not readily identifiable, and, even where it can be identified, resolution is impossible because the lost object is fundamentally irretrievable. David Eng and David Kazanjian assert that, in contrast to mourning "in which the past is declared resolved, finished, and dead," the opposite holds true for melancholy where "the past remains steadfastly alive in the present."[15] For the plastinates, the melancholic symbol of the past that persists into a present that can no longer accommodate it is life itself: the plastinates are denied the finality of death and disintegration precisely because of their imbuement with artificial life, and this forms the basis for their post-biological melancholy.

Discussing melancholy in this context does not suggest that von Hagens has deliberately fixed the plastinates in melancholic attitudes or consciously attempted to evoke a feeling of melancholy in the viewer, nor does it imply that the plastinates experience melancholy personally. Rather, the overall *affect* of the plastinate is melancholic; the plastinate as an artifact is tinged with a sense of melancholy in the same way that a photograph or painting might similarly be. The melancholy of the plastinate is thus not a universal or innate characteristic of the plastinate, but is instead situated within the realm of the imaginary; it is an imagined affect experienced by the viewer in response to the plastinate's complex ontological status, in particular its embodiment of synthetic life and the eradication of organic death.

The melancholy of the plastinates' post-biological existence is exemplified in the *London Body-Autopsied*, a figure who points with one hand to the sky and the other to the ground in a quasi-religious gesture, suggesting that ultimately all flesh returns to the earth below, while presumably, one's soul aspires to reach heaven above. Although the *London Body-Autopsied* gestures towards heaven and earth, he exists in neither of the spheres to which he points. He will not return through decomposition to the earth below, nor has he the consciousness suggested by his upward gesture. Rather, the *London Body-Autopsied* is suspended in an ontological limbo, a state somewhere between life and death in a synthetically contrived immortality. He looks longingly towards the heavens with an ascetic yearning yet will never attain the disembodied afterlife his gesture implies.

As a consequence of its liminality, the figure of the plastinate is marked by a longing for the reintegration and subsequent disintegration of the mortal body, a process prohibited by the technology of plastination. The melancholy of the plastinate pertains not simply to its embodiment of death, but rather results from the permanence of its flesh. Death has clearly taken place but cannot be resolved because it is incomplete. Eng and Kazanjian's metaphor of the continuation of the past as a kind of specter haunting the present is literalized in the case of the plastinate where the living body retains a spectral presence in the plastinated corpse as a consequence of its animated pose. The plastinate literally incorporates the living body of the past in its residue of flesh and also exists as a corpse in the present. It is through this irresolvable tension between life and death, body and corpse, that the plastinate's ontology is melancholic. The affliction of the plastinate is its post-biological immortality, which is corporeal rather than spiritual. Through plastination, the body is invested with a synthetic immortality that is permeated with a melancholic sadness over the loss of both life and death.

This melancholy that results from the impossibility of death in the plastinate's post-biological life is revealed in an analysis of *The Thinker* and the original anatomical illustration from which his pose was derived. Like many of the other plastinates in the exhibition, *The Thinker* is presented with a prop that accords with the action suggested by his pose. Yet, where the other plastinates in the exhibition are predominantly displayed with pieces of sporting equipment, the prop given to *The Thinker* is a human skull. The activity performed by this figure is thus metaphysical rather than physical; while many other plastinates pursue athletic endeavors, *The Thinker* ruminates upon a human skull comprised of blood vessels just as he is, presumably contemplating his own mortality. *The Thinker*, like the other plastinates, embodies a tension between life and death which is further indicated in the anatomical woodcut by Andreas Vesalius (fig.1) from which his pose originates. Vesalius' 1543 illustration depicts a skeleton gazing upon a skull and, like *The Thinker*, presumably contemplating its own mortality. The *memento mori* words inscribed in Latin on the plinth next to the skeleton translate as "Man's spirit lives; all else Death's hand shall claim"; the drawing is thus positioned to be an image of *memento mori* as well as one intended for anatomical education.[16] Vesalius' motto suggests that, while flesh is transient, knowledge and art are enduring; although the subject of the illustration has died and his corpse has long since decayed, his skeletal image will continue beyond death to provide viewers with anatomical knowledge.

Although *The Thinker*'s pose in Body Worlds is almost identical to that of Vesalius' figure, there is a notable difference in these two representations. Vesalius' *Thinker* is completely skeletal, as is the skull he contemplates; though animated, he is comprised of the same structural material as the symbol of death he gazes upon, thus visually reinforcing the notion suggested by the motto that death is inevitable and befalls without discrimination. In contrast, the plastinated *Thinker* has a more solid materiality than the skull he gazes upon. Whilst the skull is comprised of a semi-transparent filigree of blood vessels, revealing the transience of flesh, *The Thinker*'s body is, through the inclusion of his skeleton, given a greater corporeal weight than the symbol of death he contemplates. The skull that symbolizes death has decomposed further than *The Thinker*, and in this way, the inclusion of a body part that seems somehow "more dead" than the plastinated corpse endows the figure of *The Thinker* with a vested sense of life.

There is a peculiar circularity in the progression of the gaze in the plastinated *Thinker*. Living visitors gaze upon a plastinated human corpse (*The Thinker*), which in turn appears to gaze upon another plastinated human corpse (the skull). Through this relay of gazes, the actions of the exhibition viewer and *The Thinker* are set in parallel with one another; both of them gaze upon the dead. Consequently, *The Thinker* is both the deceased object of the visitor's gaze and becomes an active subject who gazes, subsequently playing the role of both dead artifact and living person. Through this dual embodiment, the melancholy of *The Thinker* becomes the impossibility of the *memento mori* in Vesalius' woodcut. The plastinated body of *The Thinker* may well retain a simulation of "man's spirit," but Death's hand cannot fully claim him until he is capable of decomposition.

Like *The Thinker*, *The Skin Man* also exhibits a concern with mortality and melancholy. *The Skin Man*'s "prop" is, like *The Thinker*'s, a fragment of a human body, where the "action" suggested by his pose is a contemplation of this bodily part. Yet unlike *The Thinker*, *The Skin Man* holds a fragment of himself; he looks quizzically at a complete human skin that has presumably been stripped from his own body. Like the pointing *London Body-Autopsied*, who is imbued with a sense of melancholy over the impossibility of the mortal decomposition suggested by his gesture, *The Skin Man* elicits a similar sadness pertaining to the irreconcilability of his skin and body. *The Skin Man* appears to express a posthumous longing for the impossible remarriage of his skin and flesh. This cleaving of skin from flesh is representative of the dis-integration of all the plastinates in Body Worlds: through plastination, the body is not only dissected, but its various components are inexorably separated from one another.

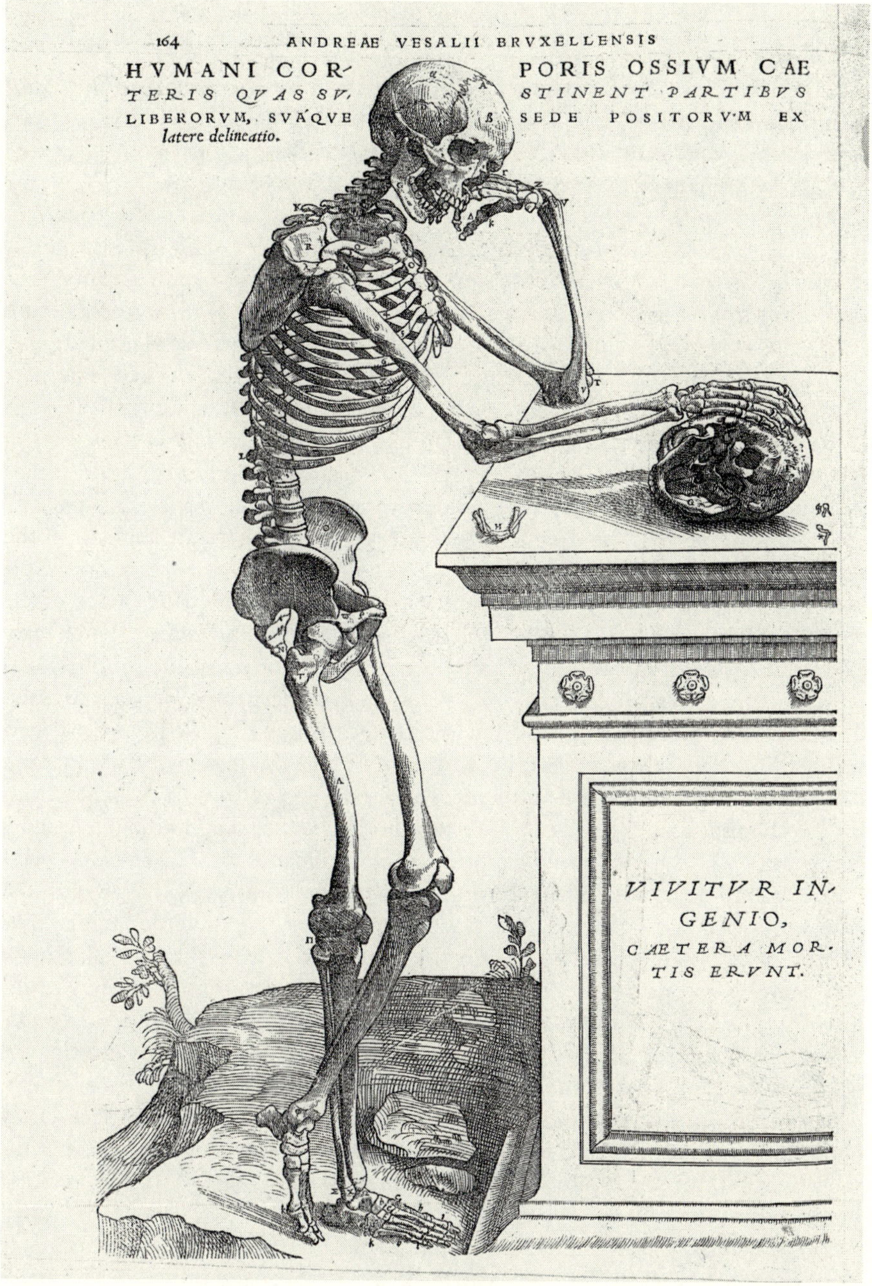

Figure 1: This woodcut, *Skeleton Contemplating a Skull* by Andreas Vesalius, from *De humani corporis fabrica libri septem* (1542), depicts a skeleton gazing upon a skull; it is the basis for von Hagens' *The Thinker* (2002). The *memento mori* words inscribed in Latin on the plinth next to the skeleton translate as "Man's spirit lives; all else Death's hand shall claim." Wellcome Library, London.

Through his pose, in which he contemplates his own body, *The Skin Man* exhibits a kind of self-referentiality that seems to explicitly acknowledge his own death and posthumous plastination. *The Skin Man*'s apparent self-awareness is contextualized by examining an anatomical illustration by Juan Valverde de Amusco from which his pose is derived (fig. 2). Like Valverde's *écorché* figure, *The Skin Man* plastinate similarly regards his skin with a kind of fascination, as if he is somewhat surprised to realize that not only has he been skinned, but also the skin he holds aloft is, in fact, his own. Both *The Skin Man* plastinate and Valverde's *écorché* figure are imbued with a sense of sorrow, as if the figures were nostalgic for the lifetime when their flesh and skin were unified. The crucial difference between *The Skin Man* and Valverde's *écorché* figure, however, is the presence of the *écorché*'s knife, which is absent in *The Skin Man.*

Margrit Shildrick discusses the significance of the knife in Valverde's illustration, noting that it is indicative of a preconception that the "self" is detachable from the flesh.[17] She describes the emergence of Cartesian dualism in the early modern period in which the body gradually came to be perceived as analogous to a machine. As Shildrick writes in relation to Valverde's image of the flayed man, the consequence of this dualism was that "in many of the early examples [of anatomical illustrations] the physical body and the motivational self are so clearly assumed to be discontinuous that the 'person' appears to assist in his own dissection. A proudly upright figure may display his own flayed skin."[18] Shildrick suggests that, in the early modern era from which Valverde's illustration originated, the body is perceived to be so profoundly detached from the psyche that one can dissect one's own body because one no longer actually embodies it. This logic initially appears to be applicable to *The Skin Man* plastinate, where it could similarly be argued that the deceased *Skin Man* played a role in his own posthumous dissection. However, unlike Valverde's figure, *The Skin Man* does not hold a knife; it is the anatomist, rather than the plastinate "himself," who is imagined to be responsible for skinning his body. Subsequently, *The Skin Man* does not have the same sense of non-corporeal agency as Valverde's figure but rather, as with all of the other Body Worlds plastinates, the "self" of *The Skin Man*'s original body has, with corporeal death, simply disappeared. In this way, the lost and irretrievable ideal that elicits melancholy is not only organic life and death, but also a mind-body dualism, and with it the possibility of spiritual resurrection, wherein subjectivity appears to exist corporeally.

This tension between conceptualizations of life and death that permeate all of the plastinates in the Body Worlds exhibition is exemplified in the figure of *The Rider*, who, like *The Skin Man*, gazes upon a fragment of his own body. Seated atop a rearing steed, *The Rider* contemplates his own brain that he cradles in an outstretched hand. His pose implies that humankind has achieved mastery over the animal kingdom through the extraordinary power of the human brain. Presumably, the "self" is somehow located within this remarkable organ,

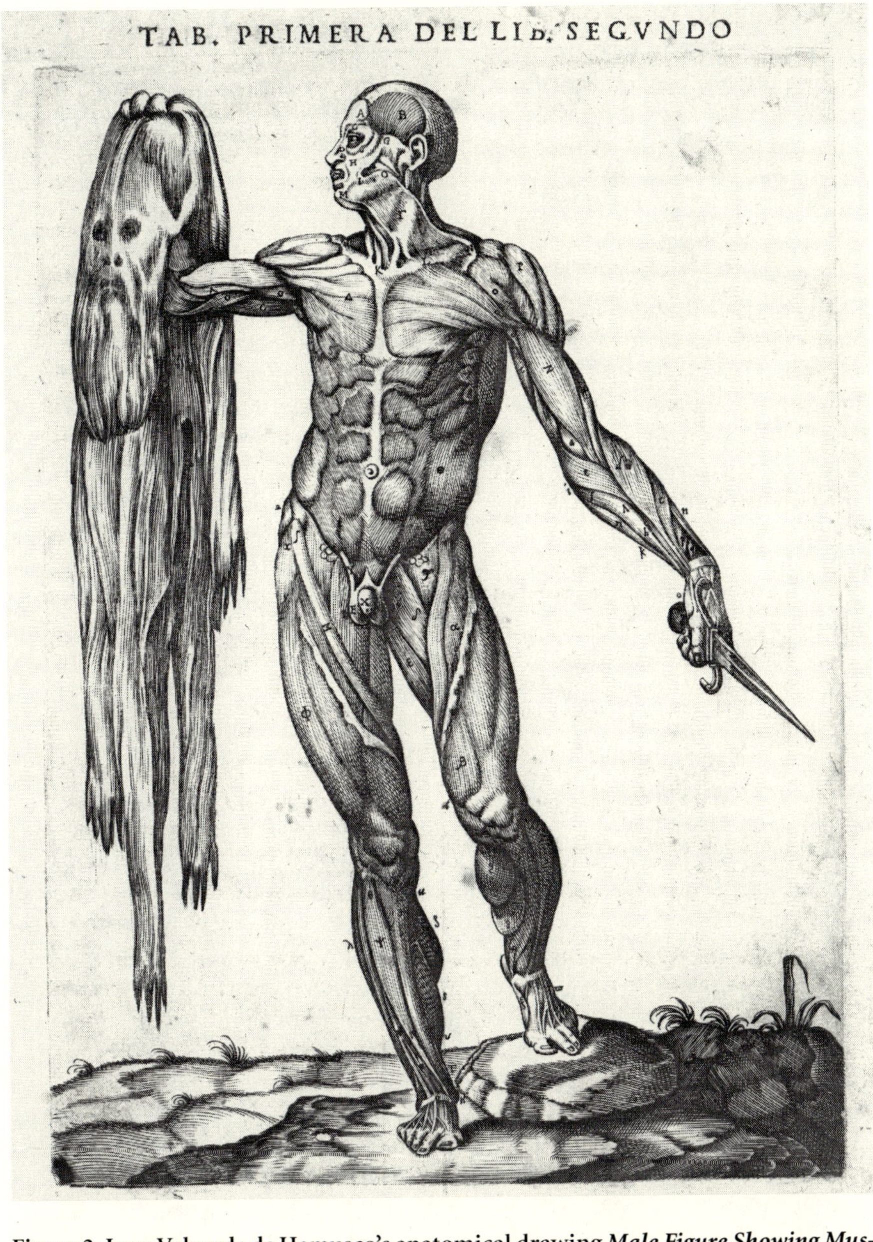

Figure 2: Juan Valverde de Hamusco's anatomical drawing ***Male Figure Showing Muscles***, from ***Historia de la composición del cuerpo humano*** (1556), is virtually identical to von Hagens' whole-body plastinate ***The Skin Man*** (1997). Notice, however, that Valverde's muscled man holds a knife, implying that he himself has participated in his own flaying, an act stressing the Cartesian duality of body and soul. Wellcome Library, London.

which acquires a fetishistic quality, symbolizing the uniqueness and evolutionary superiority of human thought. Yet, the triumph of *The Rider* over his horse and the implied dominance of the human race over the natural environment is rendered hollow by the fact that the human brain is, like *The Rider* and the horse he rides, merely reconstituted flesh. Thus there is an incongruity within *The Rider,* wherein human subjectivity, fetishized as all-powerful and transcendent, is nonetheless derived from a mortal physical artifact, one that is subject to death. Moreover, the horse too has a physical brain, one that does not appear markedly different from *The Rider*'s, thus problematizing the implicit positioning of the brain as imbued with the uniqueness of human subjectivity; after death and plastination, the distinction between human and animal is rendered void.

The hollow triumph of *The Rider*'s gesture thus pertains to the artificiality of his lifelike pose, where, despite his apparently unique human subjectivity, his carcass is like that of any other animal. As a human subject, he has ceased to exist, and his posthumous resurrection through plastination is not a spiritual rebirth but rather a synthetic immortality. *The Rider*'s situation is symptomatic of the condition of all Body Worlds plastinates, embodying competing models of corporeal and transcendent selfhood that lie in uneasy tension with one another, where the melancholy of the plastinate results from the loss of organic life and death and the loss of disembodied selfhood suggested by the exhibition's reanimation of corpses. In the case of *The Rider* plastinate, there is an unresolved ambivalence about the location of selfhood, which is paradoxically represented as both neurochemical and disembodied.

The melancholy of the plastinate is thus a consequence of the plastinate's post-biological ontology wherein the lost and longed-for object of melancholy is an organic death that synthetic resurrection denies. This sense of melancholy is in part based upon the notion of disembodied selfhood as a lost and irretrievable ideal, one that has been superseded by the simulated subjectivity of the plastinates' corporeal bodies. Within the logic according to which the exhibition operates, there is an implicit subscription to a model of disembodied subjectivity that is in tension with the exhibition's presentation of corpses. Reinvesting the plastinates with a sense of posthumous life assumes that the "self" of the once-living body no longer bears any relation to the plastinated corpse, and selfhood is thus disembodied. There is no correspondence between the posthumous identity suggested by the plastinates' actions and the identity of the living body; in this way, the body is perceived to be, as von Hagens describes, "an empty container" that, after death, may be reconstituted as an entirely new self.[19] The Body Worlds plastinates, in their post-biological existence, are thus condemned to a synthetic afterlife, unable to ever attain organic death or incorporeal resurrection.

Notes

1. von Hagens, "Anatomy and Plastination," 20.
2. Morriss-Kay, "Review: *Body Worlds,*" 536; Fischer, "When Death Goes on Display," 239; Marcus, "Birth/Death of information as art," 246; Jones, "Re-inventing Anatomy," 438.
3. Whalley, "Body Worlds," 303.
4. Ariès, *Western Attitudes Toward Death*, 89–90.
5. Ibid.
6. Ibid.
7. Blanchot, "Two Versions of the Imaginary," 81.
8. Kuppers, "Visions of Anatomy," 129.
9. Moravec, *Mind Children*, 108–109.
10. Ibid.
11. Ibid.
12. Ibid.
13. Freud, *Collected Papers*, 153.
14. Gibson, "Oublier Baudrillard: Melancholy of the Year 2000," 124.
15. Eng and Kazanjian, "Introduction: Mourning Remains," 3–4.
16. Roberts and Tomlinson, *The Fabric of the Body*, 146.
17. Shildrick, *Leaky Bodies and Boundaries*, 18.
18. Ibid.
19. von Hagens, 32–34.

Bibliography

Ariès, Philippe. *The Hour of Our Death*. Trans. Helen Weaver. London: Allen Lane, 1981.

_____. *Western Attitudes Toward Death: From the Middle Ages to the Present*. Trans. Patricia M. Ranum. Baltimore: Johns Hopkins University Press, 1974.

Blanchot, Maurice. "Two Versions of the Imaginary." In *The Gaze of Orpheus and Other Literary Essays*. Trans. Lydia Davis. New York: Station Hill, 1981.

Eng, David L., and David Kazanjian. "Introduction: Mourning Remains." In *Loss*, edited by David L. Eng and David Kazanjian, 1–28. Berkeley: University of California Press, 2003.

_____, eds. *Loss*. Berkeley: University of California Press, 2003.

Fischer, Ulrich. "When Death Goes on Display." In von Hagens and Whalley, *Gunther von Hagens'*, 236–239.

Freud, Sigmund. "Mourning and Melancholy." In *Collected Papers*, 152–170. London: Hogarth Press, 1925.

Gibson, Andrew. "Oublier Baudrillard: Melancholy of the Year 2000." *New Formations* 50, no. 1 (2003): 123–41.

Hayles, N. Katherine. *How We Became Posthuman: Virtual Bodies in Cybernetics, Literature and Informatics*. Chicago: University of Chicago Press, 1999.

Hirschauer, Stefan. "Animated Corpses: Communicating with Post Mortals in an Anatomical Exhibition." *Body & Society* 12, no. 4 (2006): 25–52.

Jones, D.G. "Re-Inventing Anatomy: The Impact of Plastination on How We See the Human Body." *Clinical Anatomy* 15 (2002): 436–40.

Kuppers, Petra. "Visions of Anatomy: Exhibitions and Dense Bodies." *differences: A Journal of Feminist Cultural Studies* 15, no. 3 (2004): 123–56.

Marcus, Aaron. "Birth/Death of Information as Art: 'Bodyworlds.'" *Information Design Journal* 11, no. 2/3 (2003): 246–51.

Moravec, Hans. *Mind Children: The Future of Robot and Human Intelligence*. Cambridge: Harvard University Press, 1988.

Morriss-Kay, Gillian. "Review: Bodyworlds." *Journal of Anatomy* 200 (2002): 535–36.

Radden, Jennifer, ed. *The Nature of Melancholy: From Aristotle to Kristeva*. New York: Oxford University Press, 2000.

Roberts, K.B., and J.D.W. Tomlinson. *The Fabric of the Body: European Traditions of Anatomical Illustration*. Oxford: Clarendon Press, 1992.

Shildrick, Margrit. *Leaky Bodies and Boundaries: Feminism, Postmodernism and (Bio)Ethics.* London: Routledge, 1997.

von Hagens, Gunther. "Anatomy and Plastination." In von Hagens and Whalley, *Gunther von Hagens'*, 9–37.

_____, and Angelina Whalley, eds. *Gunther von Hagens' Body Worlds: The Anatomical Exhibition of Real Human Bodies.* 5th ed. Heidelberg: Institut für Plastination, 2004.

Whalley, Angelina. "Body Worlds Through the Eyes of Its Visitors." In von Hagens and Whalley, *Gunther von Hagens'*, 298–303.

Locating the Sublime

Lisa Nevárez

> *Marie's eyes slammed the furthest wall after a back-forth, back-forth swinging from horror to horror, from skull to skull, beating from rib to rib, staring with hypnotic fascination at paralyzed, loveless, fleshless loins, at men made into women by evaporation, at women made into dugged swine. The fearful ricochet of vision, growing, growing, taking impetus from swollen breast to raving mouth, wall to wall, wall to wall, again, again, like a ball hurled in a game, caught in the incredible teeth, spat in a scream across the corridor to be caught in claws, lodged between thin teats, the whole standing chorus invisibly chanting the game on, on, the wild game of sight recoiling, rebounding, reshuttling on down the inconceivable procession, through a montage of erected horrors that ended finally and for all time when vision crashed against the corridor ending with one last scream from all present!*
>
> — Bradbury, "The Next in Line"[1]

In Ray Bradbury's 1947 short story, "The Next in Line," the character Marie reacts wildly to the mummies on display in the Mexican town of Guanajuato. She is particularly horrified that these bodies, which undergo a natural process of mummification thanks to the soil composition in the area, are those of individuals, Mexicans and foreigners alike, whose families cannot afford to maintain the graves and thus are put on display for visitors. Even today one can visit *las momias* of Guanajuato. Bradbury captures Marie's panicked reaction in the above quote, in which her vision spins from mummy to mummy "like a ball hurled in a game."

Bradbury's story captures a particular type of horrified reaction to being in a room full of preserved bodies, but in the twenty-first century many people are visiting museum spaces chock-full of them. This exhibit is Gunther von Hagens' Body Worlds, which has been displayed around the world since 1995, in several exhibitions. Reactions to Body Worlds often query the definition and role of art in Western culture and ask pertinent questions, such as are these plastinated bodies "art"? Where is the dividing line between scientific specimen and art object? In von Hagens' insights into his work in the exhibition catalog, *Gunther von Hagens' Body Worlds: The Anatomical Exhibition of Real Human Bodies*, either in sections he authors himself or where he is cited by critics, he reiterates that he does not present the plastinated specimens as "art" or "science," but rather aims to enlighten the viewer.[2] In this paper I do not engage in the art vs. science debate per se, but instead investigate the viewer's perception of the bodies; I am particularly intrigued by the following statistic reported after the Mannheim, Germany, exhibition (Oct. 1997–March 1998). After visitors toured the exhibition, 38 percent reported "general feelings of high intensity (either deeply moved or downcast)."[3] It is this reaction that, although devoid of Bradbury's descriptions, invites a parallel to an individual like the fictional Marie; some of the visitors to Body Worlds are disturbed to some degree by the exhibition.

One avenue for deciphering this reaction is via the sublime; what these visitors are experiencing, something terrifying but contained, may very well be that phenomenon, as explained by Edmund Burke in his 1757 work, *A Philosophical Enquiry into our Ideas of the Sublime and Beautiful.* In the following passage, Burke defines the sublime:

> Whatever is fitted in any sort to excite the ideas of pain, and danger, that is to say, whatever is in any sort terrible, or is conversant about terrible objects, or operates in a manner analogous to terror, is a source of the sublime; that is, it is productive of the strongest emotion which the mind is capable of feeling.... When danger or pain press too nearly, they are incapable of giving any delight, and are simply terrible; but at certain distances, and with certain modifications, they may be, and they are delightful, as we every day experience.[4]

Inserting an eighteenth-century treatise on the theory of the sublime, which itself follows in a long tradition of thought, seems oddly suitable for this very contemporary exhibition. Not only do the exhibition and exhibition catalog provide sociohistorical background on the dividing line between art and science and claim that the eighteenth century witnessed a divide between art and anatomical science, but throughout the exhibit viewers are reminded of the rich legacy of anatomy in art from previous eras.[5]

Inspired by this twinned emphasis on the eighteenth century and reported feelings of "high intensity," I investigate whether the Body Worlds exhibition, instead of exploring a completely new and highly controversial field of art/science, is in fact taking recourse to a much older tradition of aesthetic experience that is specifically Burkeian; I draw both on the survey results included in the exhibition catalog and on my own analyses of the plastinates. The plastinated body is particularly conducive to a sublime experience precisely because the body is stripped of most identifying features; the plastinate's sex, though, is most often obvious. The viewer looks at the specimen as an abstract figure and gazes at the interplay of muscles and bone and tissue and, in many ways, "reads" this nameless figure as a type of "landscape" or "representation" of an actual human. This seems to counter the aims of the exhibition, an observation which will be discussed later. Most importantly, Burke investigates the extremes of human emotion in his discussion of the sublime, and the fear of death plays a prominent part in that experience. The visitors' reactions from the Mannheim survey speak to a prevalent theme in the *Enquiry*, which is the continued human quest for self-preservation and acceptance of death.

In applying Burke's 1757 *Enquiry* to Body Worlds, one must remain aware that Burke's theoretical work follows in a longstanding discussion of the sublime. This begins most notably with the work of the Greek Longinus, *Peri Hupsos* or *On Sublimity*, dating from the first century, and carrying through to Despréaux Boileau's 1674 translation of Longinus' treatise. Burke's study finds itself in a tradition of eighteenth century aesthetic inquiry that his contemporary Immanuel Kant famously picks up in the *Critique of Judgment* (1790). Kant applies the sublime to the visual arts and literature, but such an analysis lies outside the scope of this paper. Analyses of the sublime continue into the twentieth century to Jean-François Lyotard in his essays, including "The Sublime and the Avant-Garde" and "After the Sublime, the State of Aesthetics." One can follow the winding path of the permutations of the sublime through various critics, but it seems fitting to examine Burke specifically for several reasons.* First, the Body Worlds exhibit avers that "the separation of art and anatomical science began by the end of the eighteenth century," which is of the same timeframe as Burke's life and work.[6] Second, the exhibit points out that von Hagens seeks a "reconvergence of science and art," which maintains a consistency with Burke's theory and his juxtaposition of that theory with examples from the physical world, such as anatomy and nature, and, although not applicable here, literature.[7] Burke's emphasis on the physical especially provides an appropriate avenue for approaching Body Worlds.

At this juncture a more detailed working definition of the sublime in Burke's treatise proves useful. The sublime, distinct from the beautiful, is associated with great size, the male sex, uneven surfaces, activity, darkness, and turbulence. The beautiful, in contrast, is identified as being female, smooth, even, passive, and light. While gazing upon the beautiful can be pleasurable and "inspire us with sentiments of tenderness," the sublime elicits more extreme emotions, such as terror.[8] An element of delight mitigates this terror, though, and the individual, while deeply moved by the experience, is not consumed by fear. Burke mentions the visual arts, and later on language, which he prefers as a means of approaching the sublime in the *Enquiry*. A painting, for example, that inspires a sublime sensation has a mimetic quality that echoes the "dark" aspects of nature that so impress us, but Burke favors literature as a way of inspiring this emotion.[9]

**Any discussion of the sublime is bound to be wide-ranging, and a more detailed history of it goes beyond the scope of this paper. Numerous critical studies exist, though, many of which examine the concept's profound impact on Romanticism. In addition to the critics mentioned above, one can certainly include Thomas Weiskel and Slavoj Zizek; for further reading on the history of critical discussion on the sublime, see Samuel H. Monk's classic 1935 study* The Sublime: A Study of Critical Theories in Eighteenth-Century England *and Philip Shaw's more recent* The Sublime.

A type of "sympathy" overwhelms the person experiencing the sublime; she could be astounded by a natural landscape or a certain sound, for example. Burke writes:

> It is by this principle chiefly that poetry, painting, and other affecting arts, transfuse their passions from one breast to another, and are often capable of grafting a delight on wretchedness, misery, and death itself ... The satisfaction has been commonly attributed, first, to the comfort we receive in considering that so melancholy a story is no more than a fiction; and next, to the contemplation of our own freedom from the evils which we see represented.[10]

The "sympathy" gives way to these sensations of "delight" and "satisfaction." The beginning of the *Enquiry* centers on many physical and anatomical examples to illustrate this theory, and Burke concludes the *Enquiry* by focusing on language and its influence on readers' experience of the sublime, such as when they read a poem and visualize the events or scene. The imagination, then, becomes central to the individual's experience.

Von Hagens' work follows in a tradition of anatomy and dissection, similar to Burke, as we will see, despite their nearly 250-year divide. The exhibition catalog, for example, takes great pains to situate "plastination" in line with Leonardo da Vinci's dissection of cadavers in von Hagens' efforts to further improve his art,[11] and continues by providing excerpts from other accounts of dissection, ranging from anatomical work in ancient Greece and Egypt to artwork from eighteenth-century France, such as Gautier's *Dissected Woman with Opened Abdomen and Fetus.*[12] Furthermore, the catalog, which I rely on as the "text" explaining von Hagens' project, also mentions at several junctures discussions of theology and ethics, drawing on Ludwig von Wittgenstein and Kant, among others. It seems clear, then, that the organizers of Body Worlds intentionally try to situate it historically and place it in a line of critical inquiry.

Despite this emphasis on the legacy of anatomy, von Hagens' work is also startlingly new. The process of plastination is itself a product of the late twentieth century, and the widespread dissemination of the exhibit, from Asia to Europe, would not be possible without contemporary means of science, technology, and transportation. The chemical process of plastination goes beyond the scope of this paper, but a description of a plastinated body will suffice, and I will draw on this particular figure in the paper: *The Soccer Player* (2005). This body is posed in the action of kicking a soccer ball, which rests on the toe of the outstretched foot. The figure balances his weight on the ball of his left foot, allowing one to see the bulging muscles of the thigh, which offset the extended right leg and extended right arm, which is thrown behind him for balance. The gaze is fixed on the ball in front of him, with the left arm curved slightly, allowing one to sense the momentum of the action. The figure is devoid of skin and hair. Instead, one can see the body in shades of red and white; and ridges and lines of muscles mark the body. The figure is clearly male; the penis and testicles are included. *The Soccer Player* resembles other plastinates in the exhibits in that they are posed in a variety of active postures and are similarly stripped of skin and most other identifying features.

Leaving *The Soccer Player* for the moment, one can turn to the visitor's reaction to seeing such a figure, and then apply Burke's theory. The first Body Worlds exhibition in Europe was located in Mannheim, and von Hagens commissioned a survey of this exhibition in the hopes that the controversy surrounding Body Worlds would become more "objective"; the initial findings of the survey were revealed in January 1998.[13] The Mannheim survey was carried out by Dr. Ernst-D. Lantermann, professor of personality and social psychology at the University of Kassel in Germany, and he writes that the goal was to obtain "the most objective picture possible of their [the visitors'] motives, expectations and fears before they had seen the exhibition, and their opinions afterwards."[14] In the catalog, Angelina Whalley echoes Lantermann's words in her discussion of the same survey: "[it] was intended to show as objective a picture as possible of the expectations, motivations and evaluations of the visitors to the exhibition as well as their resolutions for their own behavior in the future."[15]

The extremely detailed survey is divided into several sections gauging visitors' reactions at different points during their visits. Phases 1 and 2 were conducted right before visitors toured the exhibition and focused on their expectations of the exhibition, their feelings before seeing the exhibition, and how they learned about the exhibition, among other questions. Phases 3, 4, and 5 immediately followed the visit to the exhibition. Questions in those phases included asking about impressions, feelings, and aesthetics. In Phase 5, the last phase, visitors were queried about the "long-term consequences of their visit to Body Worlds."[16] While the catalog provides a general overview of surveys conducted following several other exhibitions in Europe and elsewhere, the Mannheim one is the only one quoted in detail. The survey is clearly detailed and in-depth, but one must also bear in mind when discussing the results that this is a reaction by a smaller, European population. Even the organizers of the exhibit profess their awareness of different cultural reactions. For instance, the surveyors note that visitors to the 1998 Japanese exhibition were "comparatively restrained in expressing their opinions" in contrast to European evaluations and attribute this to cultural differences.[17]

Visitors to the Mannheim exhibition totaled 777,440; 2,000 of these visitors (500 each in November and December 1997 and 1,000 in January 1998) were surveyed at random and at different times of the day and days of the week. Although the surveyors did note the age and sex, they did not divide the survey results according to that information. The survey, as mentioned previously, was based on a "five-phase" model, and allowed the surveyors to assess reactions from "prior information" in Phase 1 to reactions after the exhibit in Phase 4. The entire result of this particular question from Phase 3 follows:

> After they had been to the exhibition, the visitors were first questioned about their impressions of the various plastinates, their feelings when confronted with them, their assessments of individual groups of specimens (whole-body plastinates, isolated healthy and diseased organs, malformations, fetuses and body slices) and their opinion of their aesthetic qualities. Again the feelings expressed could be classified into three groups:
>
> 1. 64 percent of those questioned had exclusively experienced positive, information-related feelings (curiosity, amazement, fascination)
> 2. 5 percent reported exclusive feelings of rejection (having felt outrage, injured, revolted)
> 3. 38 percent reported general feelings of *high intensity (either deeply moved or downcast)*[18]

Although other critics of Body Worlds have not cited the Mannheim survey, they have identified the visitors' resulting extremes of emotions. For instance, José van Dijck, in her discussion on the "spectacle" of Body Worlds and von Hagens' style of presentation, observes that viewers have a "visceral identification with the dissected bodies."[19] One could call such a reaction "visceral," or of "high intensity." Either way, it is this passive experience of gazing at an object and then identifying a corresponding, intense emotion that possesses overtures of Burke's sublime.

Burke connects emotions such as "high intensity" to the sublime, and anatomy and nature provide examples for Burke to rely upon in the *Enquiry*. He emphasizes not only "vastness" (e.g. in nature) as a hallmark of the sublime, but also the "minute." This results when

> [w]e pursue animal life into the excessively small, and yet organized beings, that escape the nicest inquisition of the sense, when we push our discoveries yet downward, and consider those creatures so many degrees yet smaller, and the still diminishing scale of existence, in tracing which the imagination is lost as well as the sense, we become amazed and confounded at the wonders of minuteness; nor can we distinguish in its effect this extreme of littleness from the vast itself.[20]

This examination of detail as an avenue for the sublime speaks to anatomists' art, particularly in their exhibitions of the hidden minutiae of our bodies.

Reading Burke with an emphasis on science breaks with the traditional approach to the *Enquiry*; most often critics read it with an eye towards aesthetic theory or in relation to Burke's later work, *Reflections on the Revolution in France* (1790). However, Burke's connection to eighteenth-century medicine has recently been discussed by Aris Sarafianos; Sarafianos focuses on Burke's emphasis on the role of the physical body in experiencing the sublime. Sarafianos points out that Burke highlights "exercise" as key to the sublime, for example, and downplays the "relaxation" that accompanies the beautiful.[21] In addition, Sarafianos observes that Burke's "re-imagining and redefinition of health" distinguishes him from medical theorists of the time.[22] One can extend and apply such a linkage of eighteenth-century science with Burke to allow a further connection between the sublime and the plastinate.

First, two key points, divine creation and use, in the *Enquiry* prove helpful in relation to Body Worlds. In his discussion of science, Burke takes care to touch on divine creation, distinguishing between the anatomist and lay viewer:

> How different is the satisfaction of an anatomist, who discovers the use of the muscles and of the skin, the excellent contrivance of the one for the various movements of the body, and the wonderful texture of the other, at once a general covering, and at once a general outlet as well as inlet; how different is this from the affection which possesses an ordinary man at the sight of a delicate smooth skin, and all the other parts of beauty which require no investigation to be perceived? In the former case, whilst we look up to the Maker with admiration and praise, the object which causes it may be odious and distasteful; the latter very often so touches us by its power on the imagination, that we examine but little into the artifice of its contrivance; and we have need of a strong effort of our reason to disentangle our minds from the allurements of the object to a consideration of that wisdom which invented so powerful a machine.[23]

A line exists between the anatomist, who is in tune with the "Maker," and the spectator, whose reaction needs refinement. Spectators can "consider" the awe of a divine creator through their contemplation of the object. The topic of religion and ethics is certainly applicable to Body Worlds, and concerns about this have not escaped the organizers. The catalog, for example, includes several essays on ethics and the exhibition, such as those by Ulrich Fisher and Franz Josef Wetz. Connected to an appreciation of a divine creation lies the question of "use" in conjunction with aesthetics. Burke observes the following:

> The stomach, the lungs, the liver, as well as other parts, are incomparably well adapted to their purposes; yet they are far from having any beauty... It is true, that the infinitely wise and good Creator has, of his bounty, frequently joined beauty to those things which he has made useful to us [e.g. a "beautiful eye" that is used for vision]; but this does not prove that an idea of use and beauty are the same thing, or that they are any way dependent on each other.[24]

Burke distinguishes the practicality of the body's organs from their aesthetic appearance, here remarking on "beauty," which pleases the eye.

Most importantly to Burke, though, is the aforementioned role of the imagination for the individual, whether it be in visualizing a scene described in a poem or allowing the self to fully appreciate the landscape painting and be transported mentally to that site. "Expectation" feeds the imagination, which in turn contributes to a certain "tension" that factors into the sublime.[25] The visitor to Body Worlds approaches the exhibition in the same manner as Burke's reader or viewer. The former attends the exhibition, after all, by hearing about it in some way. The Mannheim survey reports that in the first surveyed group, in the first and second weeks of November, 47 percent heard about the exhibit "exclusively" through the mass media, 22 percent from friends, and 31 percent from both; by the second survey period, in early December, most had heard through friends instead of the media.[26] This indicates that visitors cannot help having had certain preconceptions or expectations which, one can conjecture, result in a type of "tension" upon entering the exhibit. After hearing reports and descriptions of plastinated bodies on display, can one expect *not* to be curious about what one will see? The Mannheim survey reveals that, prior to entering the exhibition, 86 percent had "only positive feelings (curiosity, anticipation, and expectant excitement)."[27] These sensations of "anticipation" and "excitement" certainly parallel the individual experiencing Burke's sublime.

If the link then exists between Burke and anatomy and the sublime, can one still compare a plastinate with the sublime experience in the Burkeian model? After all, gazing at a plastinate is distinct from reading a poem or viewing a waterfall because of the increased graphic quality of the plastinate and its exposed body. Integral to understanding the sublime experience in both Burke and Body Worlds is the role of separation. Burke offers, "...there is no spectacle we so eagerly pursue, as that of some uncommon and grievous calamity; so that whether the misfortune is before our eyes, or whether they are turned back to it in history, it always touches with delight ... the pain we feel, prompts us to relieve ourselves in relieving those who suffer."[28] The viewer Burke discusses has a connection with these uncomfortable sensations, but maintains a certain degree of aloofness. "Pain" operates in different contexts as applied to Body Worlds, both in terms of the visitor's "pain" upon witnessing what could be a disturbing sight, and also the concept of "pain" as projected onto the plastinate, e.g. imagining how being divested of one's skin must feel. The visitor must identify with the plastinate to a certain extent, but if gazing on a plastinate such as *The Soccer Player* were "painful," then the sublime could not occur. As Catherine Spooner argues, this identification can also be described as "uncanny": ("there was once a person like me—familiar—but now is unfamiliar—dead and dissected").[29] Although experiencing the uncanny can be an unnerving sensation, like the viewing of the plastinate, it is not terrifying to the point of petrifying the viewer.

The element of "pain" accompanies the sublime to a certain degree, but since one gazes at the plastinate with an abstraction, or distance, it is possible to view it as a *representation* of a human body. This representative aspect renders the "pain" manageable; while the visitor is fully aware that these are "real" plastinates, a certain separation is maintained. In many ways a visit to the Body Worlds exhibition *resembles* going to a museum filled with artifacts, even though many of the exhibitions do take place in such a space—the visitor buys a ticket and meanders through the displays, stopping at ones that strike his or her interest. In this sense, the visitor is predisposed to treat the exhibit as an educational experience and can gaze at a plastinate as she would a Rothko or Rossetti or an artifact from a previous era. Technically speaking, the Body Worlds visitor is seeing a museum exhibition with "real" artifacts. However, as determined by visitors' reactions in the survey and connecting them to the sublime, one is actually re-interpreting the plastinate as a *representation*. This is the avenue for experiencing the sublime. For Burke, one can experience the sublime through language precisely because one is not immediately, physically threatened. In language, such as with a poem, a "safe" distance is maintained, much like with the plastinate in the exhibition and the viewer's perception of it. The words and the bodies inspire emotion, but are not "close" enough to threaten the individual.

Van Dijck discusses the plastinate as representation also, but emphasizes that the plastinates are not representations of bodies, such as in the tradition of Renaissance artists, but rather they are "*imitations of representations*," such as the plastinate who carries his own skin, which copies Vesalius' figure.[30] I interpret her to mean that von Hagens has copied several Renaissance (and other) models in his staging of plastinates, thereby removing the viewer even further from the actual body; a semi-artistic/historical model filters the experience for some of the plastinates, hence her emphasis on "imitation." Van Dijck goes on to discuss that in the plastinate "object and representation seem to fuse."[31] In this sense, then, the plastinate may cross a line between science and art, opening up the possibility of a third category; this alternative designation is worth further exploration in another venue. Although van Dijck's discussion of imitative representation is intriguing, I argue that many of the plastinates function simply as representations, even though they are actual humans.

Since the plastinates have been divested of identifying features, such as skin, one is allowed to see them as a step removed from oneself, hence the absence of an immediate physical threat. One of the questions from the survey in Phase 5 queried the visitors on the realistic element of the exhibit; they were asked "whether their impressions and evaluations were also determined by the fact 'that real bodies and parts of bodies' were on display," and 46 percent responded "absolutely."[32] A related question asked if "an exhibition with deceptively genuine plastic models and vivid computer images would have had a similar effect," and 53 percent responded "definitely not."[33] Furthermore, visitors claim overall that they have reconsidered their attitudes towards death, viewing it in a more accepting way, and are more willing to agree to either autopsies or the plastination of their own bodies.[34] In sum, then, the fact that visitors looked at real human plastinates is vital to the success of the exhibition, as gauged by this reconsideration of death. This positive reaction could not emerge were the viewers feeling threatened, beyond the "general feelings of high intensity," by seeing the plastinates. Although the visitors are aware that the plastinates are "real" and cite that as a reason for their interest in the exhibition, there is a distance between them and the plastinate. A certain self-deception takes place at the exhibition, in which the visitors, while cognizant that they are looking at a human body, push that knowledge aside and view the plastinate from a certain emotional and intellectual distance.

Catherine Spooner echoes this non-authenticity of the plastinates, writing "[H]uman yet plastic, von Hagens' bodies are both real and a replacement of the real. They are poised on the boundary between art and science, education and entertainment, celebration and exploitation, detachment and disgust."[35] Spooner's comments echo the longstanding debate about whether Body Worlds is art exhibition or science exhibition, but also point to its position on the dividing line of both. More importantly, Spooner observes that the plastinates are simultaneously "real" and "a replacement of the real"; this discussion of an altered reality relates to the viewing experience under discussion here, as well as van Dijck's observation.

We can now turn to *The Soccer Player* as an example and look at him as a *representation* through "sublime" eyes. Devoid of skin, *The Soccer Player* is a mélange of scarlet and white, striped with ripples of striated muscle. The overall impression is one of a firm figure, and one can imagine the visitor's fingertips running along the surface, reading it like Braille. The spaces between muscles, as in the bulging thigh *The Soccer Player* uses for balance, form sudden crevices where the muscles vanish into canyons. The visitor is not permitted to touch the plastinate, but the texture is apparent. This interplay of texture and the above mentioned angles of the body of *The Soccer Player* first strike the visitor. Burke observes that while "smooth" bodies are beautiful, "[T]here can be no doubt that bodies which are rough and angular, rouse and vellicate the organs of feeling, causing a sensation of pain, which consists in the violent tension or contraction of the muscular fibres. On the contrary, the application of smooth bodies relax..."[36] In this plastinate, as in many of the others, the viewer does not confront an ovoid form, but rather angled figures who bear ridges and bumps of muscle and fiber.

This interest in and connection to the plastinate is vital to von Hagens' project. He has stated that he hopes that visitors will get a sense of the inner workings of the human body, and they can then be encouraged to take better care of themselves through viewing the exhibit. In fact, many report taking steps in that direction.[37] The text accompanying the image of *The Soccer Player* in the catalog, for instance, describes muscle contractions.[38] Von Hagens emphasizes the role of education about the body and a re-thinking of attitudes towards death in Body Worlds at several points in the catalog, including statements such as, "[L]ife is short, but Plastination will endure (and will educate) for a long time."[39] By all reports, von Hagens' educational aim has been reached in the majority of audiences. This is in part, I think, because of visitors' connections with the plastinates. They are filled with the emotions I discuss earlier, enabling them to have a profound, sublime experience when visiting a Body Worlds exhibition, and perhaps this heightened sensitivity paves the way for the "education." The sublime experience for Body Worlds' visitors goes beyond merely an appreciation for the intricacies and applications of, say, an elbow joint. One can be educated because the plastinate is a representation; visitors do not gaze at their own elbow joints, which would be terribly frightening, but rather see someone else's as analogous to their own joints, and with the conflation of emotion and anticipation the sublime appears.

In any discussion of the sublime, and especially here, one cannot separate out the gender distinction Burke pays attention to in the *Enquiry*, or its role in Body Worlds. Since the Mannheim exhibition survey does not specify the sex of the respondents, one cannot categorize the reactions of the visitors in this way; it would be interesting to know how males responded to certain questions, for instance. However, most of the plastinates—*The Soccer Player* included—are male, which according to Burke specifically characterizes the sublime. Von Hagens has stated that he intentionally included mostly male plastinated bodies in the exhibits both because of the musculature and in keeping with anatomical tradition, and that because of requests will include more female plastinates in the future.[40]

This gender distinction serves two purposes. Not only does it separate the erotic beautiful from the raw passion of the sublime, but it shuts off the feeling of "tenderness" Burke identifies with the beautiful. Even though in a Burkeian sense there is not much beauty in Body Worlds, the language throughout the exhibit and catalog relies on the term "beauty" again and again; it appears in the Mannheim survey and one learns in Phase 3 that 60 percent found "at least one exhibit particularly beautiful."[41] The word choice seems to belong to the surveyors, not the visitors, and denotes a different concept of aesthetics than in Burke's *Enquiry*. Using Burke's definitions of "beautiful" and "sublime," it is evident that the female "beautiful" does not appear in the plastinate, but rather the "male" sublime, which is privileged by Burke and capable of inspiring fear, terror, and feelings of "high intensity."

Since most of the plastinates are male, this sublime/plastinate connection fuses together. The female plastinate, however, ruptures this harmony. The plastinates remain uniform insofar as the viewer gazes at skinless forms and can then see them as representative "landscapes." In this respect, *The Soccer Player* resembles *The Archer* (2005), who is female; on both figures one can see the strips of muscle and the stark white and red of bone and tissue. *The Archer* is posed in arguably a gender-neutral position, like some of her fellow plastinates. More typically, though, the plastinates are posed in positions that emphasize a gender stereotype, such as *The Ballet Dancer* (undated), who wears pink *pointe* shoes, or *The Swordsman* (1999), who is posed in a lunge that clearly reveals his male sex organs. Should the exhibition expand to include more female bodies, in more "female" positions, one can anticipate a different type of sublime — or perhaps beautiful — reaction. As it is, the exhibit does contain displays of pregnant women, such as *Woman Bearing Life* (1996) and *Reclining Pregnant Woman* (1999). The latter prompted a range of reactions, one from a woman (self-identified as being a mother) who stated "I have first truly understood the miracle of pregnancy while in front of this woman's body" to another who deplored the "pin-up pose" as "not being in very good taste."[42]

With plastinates such as these, the sublime reaction is problematized (if it can even be said to exist) in comparison to that prompted by the male plastinates.

The female plastinate in a "female" position may inspire the sublime in the same way gazing at a male plastinate would, with the same exposed muscle and tissue, but concurrently one is startlingly aware of her sex. The smooth curve of the pregnant belly and the gentle oval of the fetus curled in a fetal position inside the plastinated mother break from sublime aesthetics. Burke writes about beautiful female bodies in the *Enquiry*:

> But as perfectly beautiful bodies are not composed of angular parts, so their parts never continue long in the same right line... Observe that part of a beautiful woman where she is perhaps the most beautiful, about the neck and breasts; the smoothness; the softness; the easy and insensible swell; the variety of the surface, which is never for the smallest space the same; the deceitful maze, through which the unsteady eye slides giddily, without knowing where to fix, or whither it is carried.[43]

Burke rhapsodizes about the smooth curves of the female body as illustrative of the beautiful. The two pregnant female figures mentioned above fit into Burke's description of smooth, ovoid, "beautiful" parts, with the mound of the pregnant belly and the swell of the breasts. There are few female plastinates in poses such as these, so the overwhelming sublime masculinity of Body Worlds still emerges, but—should the exhibition change to include more women in more "female" positions—corresponding surveys addressing the role of gender and sex would be fascinating.

Others would disagree that one can trace the sublime in a plastinated body at all, regardless of the sex. Jonathan Padley, for instance, in his article contrasting von Hagens with Dr. Frankenstein of Mary Shelley's *Frankenstein* (1818), argues that there cannot be a sublime in Body Worlds because the plastinated bodies "do not impact irrevocably upon the ongoing processes of physical creation, and thus they cannot impose a consummately terrifying effect. Their respective sublimities are limited, and merely arise from the fact that they are unnerving to some of their observers."[44] I would disagree. While Padley is correct insofar as these plastinates are "mechanical" because they are no longer alive, I would say that one can react just as powerfully to them as to a scene in nature, and thereby can experience the sublime. As the Mannheim survey reveals, "unnerving," or being unnerved, can be a powerful sensation.

Padley is not alone in seeing these plastinates as inert; van Dijck, although focusing on the visitors' artistic/scientific reactions, observes that the plastinates are hybrid and "fake" in that they have been manipulated, and that it is a mark of our postmodern culture that we cannot readily distinguish between a "body" and a "model."[45] The conversation here can veer towards a discussion of hyperreality, to use Jean Baudrillard's term, that so distinguishes our postmodern world. Despite the overtones of the postmodern, the visitors' reactions follow in a similar vein to Renaissance reactions to dissections, during which viewers were urged to "identify" with the dissected body so as to relate to their own futures.[46] The Body Worlds visitors do something similar, as the Mannheim survey reveals, in that many leave with a renewed awareness of their own health. As van Dijck points out, though, the Body Worlds organizers capitalize on this attention to one's body and more positive attitude towards plastination by offering forms at the end of the exhibit so the visitors can sign up as future donors.[47]

Veiled with scientific terminology, and often wedged into an art vs. science debate, the Body Worlds exhibition is, I contend, positioned in a tradition of sublime aesthetic inquiry that emphasizes the role of the viewer and her sensory experience, the role of representation, the distinction between male and female, and attitude towards pain. By following a Burkeian reading, one can position Body Worlds in a larger tradition and also better comprehend what viewers report feeling upon leaving the exhibition. Aside from the ongoing and thought-provoking debate over its role as art or science, a Burkeian reading may be one way of addressing the controversy this exhibition has provoked around the world and explain why the project fascinates people. Although Marie in "The Next in Line" finds herself completely unnerved upon entering a room containing preserved bodies, visitors to Body Worlds may not find themselves too disconcerted upon visiting a museum full of plastinates. The Body Worlds visitor, after all, may receive something else with the price of admission: a sublime experience.

Notes

1. Bradbury, "The Next in Line," 395.
2. Quoted in Bauer, "Plastinated Specimens," 229.
3. Lantermann, "Körperwelten," 307.
4. Burke, *A Philosophical Enquiry*, 36–7.
5. Moore and Brown, "Gunther von Hagens," 212.
6. Ibid.
7. Ibid.
8. Burke, 39.
9. Ibid., 58–9.
10. Ibid., 41.
11. Kriz, "Foreword," 6.
12. Wetz, "Dignity," 241.
13. Lantermann, 304.
14. Ibid.
15. Whalley, "Body Worlds," 299.
16. Lantermann, 309.
17. Whalley, 299–300.
18. Lantermann, 307 (emphasis mine).
19. van Dijck, *The Transparent Body*, 58.
20. Burke, 66.
21. Sarafianos, "Burke's Physiological Iconography," 229.
22. Ibid., 233.
23. Burke, 97–8.
24. Ibid., 97.
25. Ibid., 127.
26. Lantermann, 306.
27. Ibid.

28. Burke, 43.
29. Spooner, *Contemporary Gothic*, 63.
30. van Dijck, 53.
31. Ibid., 54.
32. Lantermann, 307.
33. Ibid.
34. Whalley, 303.
35. Spooner, 65.
36. Burke, 137.
37. Whalley, 303.
38. von Hagens and Whalley, 164.
39. von Hagens, "Anatomy and Plastination," 35.
40. Institute for Plastination, "Questions & Answers."
41. Lantermann, 307.
42. von Hagens, "On Gruesome Corpses," 264.
43. Burke, 104–5.
44. Padley, "*Frankenstein* and (Sublime) Creation," 203.
45. van Dijck, 62.
46. Ibid., 59.
47. Ibid.

Bibliography

Bauer, Axel W. "Plastinated Specimens and Their Presentation in Museums: A Theoretical and Bioethical Retrospective on a Media Event." In von Hagens and Whalley, *Gunther von Hagens'*, 222–34.

Bradbury, Ray. "The Next in Line." *The Stories of Ray Bradbury*. New York: Knopf, 1980.

Burke, Edmund. *A Philosophical Enquiry into the Origin of Our Ideas of the Sublime and Beautiful*, edited by Adam Phillips. Oxford: Oxford University Press, 1990.

Institute for Plastination. "Questions & Answers." Gunther von Hagens' Body Worlds: The Original Exhibition of Real Human Bodies. Heidelberg: 2006. http://www.bodyworlds.com/en/exhibitions/questions_answers.html (accessed June 9, 2007).

Kriz, Wilhelm. "Foreword." In von Hagens and Whalley, *Gunther von Hagens'*, 5–8.

Lantermann, Ernst D. "Körperwelten as Seen by Visitors; Visitor Survey Report from the First Exhibition in Europe." In von Hagens and Whalley, *Gunther von Hagens'*, 304–11.

Monk, Samuel H., *The Sublime: A Study of Critical Theories in Eighteenth-Century England*. New York: MLA, 1960.

Moore, Charleen M., and C. Mackenzie Brown. "Gunther von Hagens and Body Worlds." In von Hagens and Whalley, *Gunther von Hagens'*, 208–21.

Padley, Jonathan. "*Frankenstein* and (Sublime) Creation." *Romanticism* 9, no.2 (2003): 196–212.

Sarafianos, Aris. "Burke's Physiological Iconography of Aesthetic Perception and the Invention of Sublime Medicine." *Comparative Critical Studies* 2, no. 2 (2005): 227–40.

Shaw, Philip. *The Sublime*. London: Routledge, 2006.

Spooner, Catherine. *Contemporary Gothic*. London: Reaktion, 2006.

van Dijck, José. *The Transparent Body: A Cultural Analysis of Medical Imaging*. Seattle: University of Washington Press, 2005.

von Hagens, Gunther. "Anatomy and Plastination." In von Hagens and Whalley, *Gunther von Hagens'*, 9–37.

_____."On Gruesome Corpses, Gestalt Plastinates and Mandatory Interment." In von Hagens and Whalley, *Gunther von Hagens'*, 260–82.

_____, and Angelina Whalley, eds. *Gunther von Hagens' Body Worlds: The Anatomical Exhibition of Real Human Bodies*. Translated by Francis Kelly. Heidelburg: Arts & Sciences, 2007.

Weiskel, Thomas. *The Romantic Sublime: Studies in the Structure and Psychology of Transcendence*. Baltimore: Johns Hopkins University Press, 1976.

Wetz, Franz Josef. "The Dignity of the Individual." In von Hagens and Whalley, *Gunther von Hagens',* 241–59.
Whalley, Angelina. "Body Worlds through the Eyes of Its Visitors." In von Hagens and Whalley, *Gunther von Hagens',* 298–303.
Zizek, Slavoj. *The Sublime Object of Ideology*. New York: Verso, 1989.

Take Me: The Rhetoric of Donation

Aviva Briefel

> *You've "cut through" a taboo, as it were, and for that I'm particularly grateful to you. And there's one thing I'm now sure of: when I've breathed my last, my body will perform an important function for mankind. Mind and spirit will gaze upon it with envy... I even dreamed that I had been cut into fine slices and was admiring myself. It was a lovely dream.*[1]
>
> —Testimony from body donor

What is appealing about imagining one's own anatomized body? The donor transforms a grotesque nightmare into an enticing dream; the horror of bodily fragmentation gives way to narcissism. This reaction is striking given the gothic associations that seem to pervade the Body Worlds exhibits. As one critic comments, the plastinates resemble "zombies as portrayed in the movies, with eyes that stare out impassively and with strips of skin ... and sporadic hair framing the raw red and white muscles of the face."[2] It is tempting to compare Gunther von Hagens' life (and death) work to the pantheon of mad scientists and monsters that pervade gothic literary and cinematic texts. The image of Doctor Frankenstein readily comes to mind, as does Wes Craven's Freddy Krueger, whose flayed features and habit of wearing a dark fedora combine the appearances of von Hagens and his plastinates. Or else, the exhibit may recall Vincent Price's B-movie *House of Wax* (1953, dir. André de Toth), in which a deranged curator revamps his wax museum by showcasing real corpses; as he aptly explains, "Each subject must be taken from life." The artistic positioning of the plastinates, their presentation of human body parts in sculptural forms, also evokes the grotesque statues composed of human remains in Tobe Hooper's *The Texas Chain Saw Massacre* (1974). The list of gruesome resonances goes on.

And yet, for many of its visitors, Body Worlds is not only antithetical to the gothic, but also staves off its dark implications. In the words of one donor, "The thought of being buried in the earth after my death is a horrible one as far as I'm concerned, because I am completely disgusted by worms and grow panicky when I think of them. Now I can relax about the issue, though, since

I can stay above ground after my death and may even have a chance to 'experience' trips (i.e. exhibitions) to many different places."[3] As evidenced in the enthusiastic tone of this excerpt, plastination may serve as a remedy for the oblivion and fear associated with death. The donor celebrates her ability to replace morbid phobias with the excitement of postmortem travel. According to another donor, "Ever since I heard about the plastination process, the prospect of my own death hasn't bothered me at all."[4] Rather than being a chamber of horrors, Body Worlds may offer eternal life.

In his various writings, von Hagens adamantly asserts that his exhibits are neither morbid nor gothic. He explains that the only shock viewers should receive from them is "an aesthetic shock, not a cruel one. Hollywood has done a lot to associate anatomy with the horror and emotional distress of dying. But my work is very different."[5] This "difference" is based partly in his attempts to discourage audiences from considering the plastinates repulsive; these human statues are sanitary, odorless, and arranged in pleasant settings intended to evoke natural landscapes rather than lugubrious crypts.[6] Most obviously, however, von Hagens combats allegations of horror with an explicit rhetoric of donation that promises full agency to the potential donor. This logic rests on the assumption that volunteerism and free will are antithetical to victimization and violence. In what follows, I will suggest that volition is not foreign to the gothic, but may indeed be one of its primary aspects. Body Worlds perpetuates a version of the gothic that purposefully blurs the line between victimhood and voluntary submission.

Again and again, von Hagens assures us that his plastinates are not victims. His official writings reiterate the strict rules followed by the Institute for Plastination (IfP) in acquiring corpses: none are obtained through coercive or illicit means; the specimens must have "died of natural causes"; the institute has "never used unclaimed or found bodies, or bodies of executed Chinese prisoners, in its anatomical exhibitions of real human bodies"; and von Hagens "ordered that all human specimens with head injuries of any kind be sorted out and cremated."[7] While several of these assurances respond to allegations about the unethical sources of certain plastinates, they also acknowledge the unease that may come from seeing corpses (whatever the origin) exhibited in a museum. The implication behind von Hagens' various reassurances is that we should neither fear nor be repulsed by these bodies, precisely because none of them was acquired through coercive means. As he explains in one of his essays, "Introducing a formal donation policy for plastination reflects the conviction that the values of a democratic society, i.e., that individuals have as much discretion over the use of their bodies as possible, dictate that a public exhibition can only be based on a conscious decision on the part of donors."[8] The presence of the plastinates proves that their donors exerted a democratic prerogative by accepting the scientist's open invitation to become a part of Body Worlds. Significantly, von Hagens delivers this summons from the perspective of someone who will also become a plastinate once he dies: "I would be happy to welcome you in the circle of donors to which I also belong."[9]

The donation brochure, which can be downloaded from the Body Worlds website or picked up at the exhibition, enforces an ideology of consent. Von Hagens clarifies that this form is not an official "*contract*," but "*merely a declaration of intent*"[10] — it is an illocutionary act that constructs the donor into a willing subject. The content and structure of the brochure invite readers to pass through a series of gateways, which assert the notion of consent in increasingly emphatic ways. The form begins with four enumerated points, the first of which instructs us to "read the information brochure 'Donating Your Body for Plastination' carefully." The imperative formulation of the statement captures the tenor of the entire document. Paradoxically, the statement commands us to follow an order that will bolster our free choice; we *must* read the pamphlet to have full knowledge of what we are getting ourselves into. The second point sets up hypothetical opponents to the donors' wishes, only to reaffirm the extent to which Body Worlds grants us agency. The potential opponents here are the donors' families: "If your relatives object to your body donation for Plastination, you may wish to also have a lawyer witness your signature on both forms." The institute and donors are now aligned against the intrusion by a third party. The next statement takes the form of an imperative that once again cedes to free choice: we are told that we "should carry [the laminated donation] card with [us] at all times." The card becomes a marker of the donors' identities and another proof of agency; our wishes will be respected if we are in possession of this document at the time of our deaths. Finally, the last point solidifies the readers' authority by confirming our right of refusal: "You may withdraw your consent to donate your body at any time, by sending us a letter, signed by you, stating that you have withdrawn your consent. You do not need to provide any reason for withdrawing your consent. Your consent will be deemed withdrawn as soon as we receive your signed letter." This repetitive assurance of sanctioned withdrawal uses the decision *not* to do something as a guarantee of the freedom inherent in doing it.[11]

The central part of the form — the heart of the consent-granting process — transforms donors into speaking subjects. The diction shifts to the first person as the document verbalizes our desires: "I do not wish to be cremated or buried. I am making this gift freely and voluntarily, without obligation of any kind on the part of the recipient organization, and there will be no reward or compensation to my family or me." The institute speaks for us, but not coercively; it merely restates our full authority. Like the previous section, this one ends with the possibility that donors may change their minds, but this idea is now stated from the donors' perspective: "I may withdraw my consent for my body to be used for Plastination at any time without having to give a reason...." Interestingly, this is followed by the assertion of one way in which the Institute for Plastination can assert control over donors' bodies — by rejecting them: "The IfP also has the right to issue a statement withdrawing its agreement to accept a body for Plastination."[12] The donors and the IfP share the power of refusal, a power that prevents the forced appropriation of our corpses. At the same time, this statement protects the reader from a fear of rejection by veiling the language of intimacy that has dominated this part of the form; it vaguely refers to "a" body rather than to "your" body.

The final section of the form emphatically confirms our free will and individual identities. It begins with a reference to the "wishes of the body donors," reconfiguring consent as a personalized desire rather than a mere statement of agreement. What is particularly telling about this section, however, is that it *follows* rather than precedes the consensual signatures of the middle section; it is thus non-binding. The second section had attenuated this fact by once again playing on the readers' consent: "I agree that my preferences as indicated in the following questions are recommendations rather than binding terms." In essence, donors willingly agree to negate the authority of the ensuing statements. This is not evident in the questions themselves, which use a contractual language while subtly omitting the official guarantee enforced by a contract. Indeed, many of these stipulations seem to grant donors a sense of postmortem control that — by its very definition — can never be ensured. Note the following statement: "Plastinated specimens, especially whole-body plastinates, are occasionally interpreted as anatomical works of art. Hence, I agree that my body can be used for an anatomical work of art." Whereas the first part of the statement refers to a hypothetical museum visitor's interpretation (whether or not something is "interpreted" as an artwork depends on the eye of the beholder), the second part emphasizes the deceased donors' agency. How can donors control the reception of their corpses? This statement grants the illusion of agency to a circumstance in which it is not possible. We find a similar dilemma in the statement that reads, "When whole-body plastinates are exhibited in public, people are always amazed and slightly dubious about the realism of the specimens. On various occasions, people have expressed a wish to be able to touch the specimens. I agree that laypeople be allowed to touch my plastinated body." While certain Body Worlds displays are meant to be handled, it is difficult to guarantee (as I witnessed during my visit to Body Worlds 2 in Montreal) that visitors will refrain from touching the cordoned-off exhibits. The sense of an overblown, quasi-supernatural consensus constructs subjects who can will their fates indefinitely. The section ends with a confirmation of personal agency by asking donors to explain their reasons for agreeing to become plastinates ("I would like to donate my body to serve a good cause," "I would like to contribute to medical research," "I am fascinated by the thought of being preserved forever for posterity," etc.). It is telling that this question should be included with the falsely consensual statements, since it does not ask donors to agree to anything; instead, they must simply assert themselves as desiring subjects.[13]

The rhetoric of consent that dominates Body Worlds allays some viewers' worries about the morbid or coercive implications of the exhibit. As philosophy professor Franz Josef Wetz explains in his catalog essay, "It must be concluded that who-ever, while he is alive and in full possession of his mental faculties, voluntarily and after comprehensive information makes his body available for this procedure, and gives his consent to exhibition of the denaturized and transformed dissection resulting from it, is not offending the dignity of his body. In contrast, whoever forbids such a donation of a body is infringing the dignity of the deceased."[14] Another writer asserts that as opposed to clear cases of human rights violations, including "the work of Anthony-Noel Kelly, the artist imprisoned for theft of body parts for use in his sculpture, ... the [Body Worlds] 'participants' gave permission for their bodies to be preserved, dissected and displayed. Under such circumstances, whatever people may think of the result, they have no cause to complain. Those who prefer not to see the exhibition need not go."[15] These two arguments foreground consent as the primary determinant of the ethics of the exhibit; both adopt a similarly dismissive stance toward those who might question it. Likewise, Lawrence Burns proposes that one way of rectifying what he views to be the problematic anonymity of the exhibits—that the displays bear von Hagens' signature but not the donors' names—would be to place greater emphasis on the donors' consent: "Perhaps their voice could be broadcast reading a detailed letter indicating why they donated their bodies and why they wish to be posed in the particular poses in which they have been placed." Burns adds that these voices need not necessarily belong to the subjects themselves, thereby confirming the value he and others place on consent alone.[16]

Megan Stern illuminates the importance of consent to Body Worlds by explaining that the exhibits need to be understood within a consumerist context. She writes, "That seeing the plastinates might prompt the desire to become one is clearly very far from the abject revulsion which Frankenstein's creature prompts.... The plastination donor, I would argue, is represented as an involved, independently minded consumer, making positive choices about his or her own death." She contrasts this to the established discourse of organ donation, which tends to prioritize the importance of the donor's gift to society rather than the benefits of donation to the self.[17] Given that consumer choice is one of the principal ways in which we come to define ourselves in modern society, the palliative nature of this discourse makes sense. Of course, as Stern convincingly suggests, the exhibits must repress the fact that their subjects essentially agree to become commodities themselves, thus illustrating "the extent to which consumer fantasies of individual empowerment and fulfillment are bound up with the problematic commodification of the body."[18] Certain donors reflect this paradoxical situation by trying to promote themselves: "My jaws could be especially interesting. The teeth were never straightened and I had an infection in the lower one on the left."[19] Body Worlds seeks to forestall the donors' sense of objectification by perpetuating the illusion that choice continues after death, that subjects will be able to control whether they are perceived as art objects or scientific specimens.

But does consent necessarily produce reassurance? Can it really assuage gothic fears associated with mad scientists and violated bodies? The issue of consent and its relationship to victimization was foregrounded in a gruesome German murder case that reached international headlines in 2002, during the same period in which the Body Worlds controversy was being discussed worldwide. A computer programmer, Armin Meiwes, had placed a personal ad on the internet seeking for a volunteer to be slaughtered and consumed by him. The ad was answered by Bernd-Jürgen Brandes, who offered himself as Meiwes' victim. One of the most shocking aspects of the case was that Brandes *fully consented* to his fate at various points during his ordeal. He responded to the ad, composed his will, stated his understanding of what was going to happen to him in a testimonial videotaped by Meiwes, and assented to the different steps leading to his demise (such as, for instance, having Meiwes cut off and cook his penis so that they could both consume it). One of the main issues raised in the courtroom was whether Meiwes could be tried for premeditated murder rather than euthanasia, given that Brandes' consent might reveal that their interaction was one of assisted suicide (Meiwes ultimately received a life sentence). In this case, the victim's consent added to — rather than detracted from — the disturbingly gothic aspects of the transaction. Inevitably, journalistic accounts centered on Brandes' active acceptance of his fate as one of the most disturbing features of the story; as one newspaper put it, this aspect is "what really chilled the bones," while one headline sensationally announced, "Victim of Cannibal Agreed to be Eaten."[20] The victim's consent aggravated the violence of the murder.*

For reasons that are too obvious to enumerate, this cannibalism incident is factually and ethically different from Body Worlds. However, the way in which consent added to the horror of the case complicates the disassociation of consent from victimization that is central to the exhibits. It demonstrates that victimhood can entail free agency, that these are not mutually exclusive terms. If we turn from real-life scenarios to the world of fiction, we find that consent has long been an important component of the gothic literary mode. In Bram Stoker's *Dracula* (1897), for instance, the infamous vampire lures his soon-to-be victim Jonathan Harker into his deadly castle by asking for his consent:

> "Welcome to my house! Enter freely and of your own will!" He made no motion of stepping to meet me, but stood like a statue, as though his gesture of welcome had fixed him into stone. The instant, however, that I had stepped over the threshold, he moved impulsively forward, and holding out his hand grasped mine with a strength which made me wince, an effect which was not lessened by the

**For summaries of the case, see Landler, "Eating People"; "German Cannibal Victim's Will"; Naughton, "German Cannibal Jailed." For a philosophical analysis, see Wisnewski, "Murder, Cannibalism, and Indirect Suicide."*

> fact that it seemed as cold as ice—more like the hand of a dead than a living man.[21]

Voluntary choice is the first step in what will become an undeniably coercive interaction; after he is wined and dined by the Count, Harker finds himself imprisoned in the castle, which he leaves through a precarious flight. As soon as he encounters the vampire, free will abruptly transforms into its opposite. The scene may be read as an allegory for the structure of the gothic novel, which proffers an open invitation to the reader (the consumer of the book) that ends as narrative imprisonment. While we are, of course, free to abandon the novel at any time, it locks us into a narrative form from which we cannot escape.

Gothic narratives frequently deploy framing devices that—like the labyrinthine spaces that are endemic to the genre—force us to become permanent fixtures within the story. These narratives position us alongside diegetic witnesses, readers, or listeners of gothic stories, only to move us to the center of the horrific events. Henry James' *The Turn of the Screw* (1897), a novel that the author famously described as a "trap for the unwary," begins by placing us within a fireside group of characters listening to ghost stories: "The story had held us, round the fire, sufficiently breathless...."[22] However, James prevents us from maintaining the safe distance of this external frame by ultimately trapping us within the abyss of its central ghost story. At the end of the text, we are left alone with the governess, the embedded narrator of this story, who proves to be more dangerous than the specters she claims to see. Instead of taking us back to the comfortable circle by the fire, the narrative abruptly concludes with the governess' murder of one of her young charges: "We were alone with the quiet day, and his little heart, dispossessed, had stopped."[23] This violent ending leaves us with nowhere to go; the inclusive "we" of the quote seems to suggest our complicity as direct witnesses of the governess' act. James borrows this claustrophobic structure from earlier gothic texts, including Mary Shelley's *Frankenstein* (1818) and Robert Louis Stevenson's *The Strange Case of Dr. Jekyll and Mr. Hyde* (1886). By asking us to identify with characters who encounter the story through written or oral accounts, the works lure us into narrative *mises en abyme,* in which we must trade our initial position as distanced spectators for our direct involvement and entrapment. In Stevenson's novel, for instance, we begin as the companions of two respectable men strolling through London and discussing the strange story of Mr. Hyde, and we conclude the text in the presence of the monster (and, inevitably, his creator) through our reading of Jekyll's suicide note.[24] At the end of Shelley's text, we are stranded on board a ship in the North Pole, just having witnessed the death of Dr. Frankenstein and the escape of his creature: "He was soon borne away by the waves, and lost in darkness and distance."[25] In each of these situations, we are marooned in a position of narrative indeterminacy, passengers on a sinking ship.

This structure compels the gothic reader to associate with the victims of the story, who have also become entrapped in the clutches of the narrative and its monsters. Whether the victim is an eighteenth-century heroine stranded in an abandoned castle with a mad aristocrat, or one of Dracula's captives, the primary threat generated by the gothic is entrapment. The act of reading emulates this structure by transforming a process that confirms free agency (the privileges of education and discrimination that allow us to read a book of our choice) into a coercive one. The same can be said of modern horror films, which capture viewers in their narrative structures to parallel the endangerment of their characters. At the end of Wes Craven's original *Nightmare on Elm Street* (1984), for instance, we find ourselves in the same indeterminate position as the onscreen teenagers. Like them, we are ensnared in a dream out of which Freddy will tirelessly emerge. The structure of the contemporary horror film — which calls for continuity through the sequel — never allows us to depart from the narrative world: Freddy, clad in his filthy fedora, is always out there, ready to return.

Despite its apparent rejection of gothic sensationalism, Body Worlds reproduces this uncanny structure. Ideal visitors take on the role of active consumers by purchasing a (rather expensive) ticket and entering the exhibition of their own free will. The exhibit gradually eases us into its depths, beginning with benign skeletons (we have all seen those before) and slowly drawing us toward the full-body plastinates. We are made aware of our volition once again by having to decide whether we want to view truly "controversial" material, such as the exhibit on reproduction concealed behind a curtain, to be entered at the viewers' discretion. After making our way along a series of anonymous corpses that are preserved, sliced, and posed, we approach the exit, which offers a return to the world of the living. However, access to this exit is intercepted by the donation booth located toward the end of the exhibit, which offers visitors the opportunity to become permanent parts of what they have just seen. The booth is the museological equivalent of the incomplete frame of the gothic narrative, which allows readers free entry but no exit. In the words of one critic, "A delicious frisson of personal death animates this little display. If life has become a little too abstract in the specimens, something that is grasped in its hygienic, clean, structured stage, then to be addressed as a person who will die seems an exciting contrast."[26] The exhibit appeals to the viewers' volition, while envisioning a period in which this volition will be taken away through our deaths. While, clearly, the choice to stay is our own rather than that of a sinister plotter, the structure of enlistment and permanent belonging bears an unmistakably gothic mark.

I want to end by suggesting that this similarity is not coincidental. Rather than rejecting the gothic altogether, Body Worlds produces a version of the genre that purposefully blurs the line between victimhood and voluntary submission. This strain of the gothic allows Body Worlds to maintain its appeal to free agency, education, and well-being, while conveying a sensational structure of gothic showmanship. Visitors experience the thrill of knowing that they may voluntarily trap themselves in this exhibit, that one day they will appear among the aesthetically posed bodies on display. This gothic tension is reflected in the whole body plastinates themselves, which convey both the active agency of the living and the permanent passivity of the dead. They are skiers, ballerinas, swimmers, horseback riders, etc., who appear to partake enthusiastically in activities that they may never have practiced as living beings.[27] Some of them seem to flaunt their flayed skin proudly or to display the organs exposed in drawers on their bodies, creating a peculiar tension between what is voluntary and what has been forcibly imposed. The focus on donation that pervades Body Worlds gives us the sense that the bodies' original owners actively consented to becoming parts of the exhibit, but that their intentionality has been transformed into something else — the visual display of the curator's vision of what volition looks like. This complication of the categories of agency and victimhood lies at the core of the sensationalism of Body Worlds, a spectacle that may herald many more "real-death" models of gothic participation.

Notes

1. von Hagens, "Anatomy and Plastination," 31.
2. Burns, "Gunther von Hagens' BODY WORLDS: Selling Beautiful Education," 14.
3. von Hagens and Whalley, *Donating Your Body for Plastination*, 27.
4. von Hagens, "Anatomy and Plastination," 30.
5. von Hagens, "Why I Think It Is High Time We Democratised the Science of Anatomy," 18.
6. von Hagens, "Anatomy and Plastination," 20. See also von Hagens, "On Gruesome Corpses, Gestalt Plastinates and Mandatory Interment," 266, 269.
7. von Hagens and Whalley, 28; Institute for Plastination, "Original and Copycat"; Goldman, "Goriest Show on Earth."
8. von Hagens, "On Gruesome Corpses," 278.
9. Hirschauer, "Animated Corpses: Communicating with Post Mortals in an Anatomical Exhibition," 35.
10. von Hagens and Whalley, 27.
11. Institute for Plastination, *Body Donation for Plastination*, 1.
12. Ibid., 2.
13. Ibid., 2–4.
14. Wetz, "The Dignity of the Individual," 254.
15. Pollard, "Our Lives, Our Bodies: Thorny Issues of Consent," 172.
16. Burns, 21.
17. Stern, "Dystopian Anxieties versus Utopian Ideals: Medicine from *Frankenstein* to *The Visible Human Project* and *Body Worlds*," 78.
18. Ibid., 81.
19. Hirschauer, 34.
20. Deutsch Welle, "German Cannibal Trial"; Harding, "Victim of Cannibal Agreed to Be Eaten."
21. Stoker, *Dracula*, 15–16.
22. Edel, *Henry James*, 29; James, *Turn of the Screw*, 21.
23. James, 116.
24. Stevenson, *The Strange Case of Dr. Jekyll and Mr. Hyde*, 54.

25. Shelley, *Frankenstein*, 166.
26. Kuppers, "Visions of Anatomy: Exhibitions and Dense Bodies," 140.
27. Walter, "Body Worlds: Clinical Detachment and Anatomical Awe," 469.

Bibliography

Burns, Lawrence. "Gunther von Hagens' BODY WORLDS: Selling Beautiful Education." *American Journal of Bioethics* 7, no. 4 (2007): 12–23.

Edel, Leon. *Henry James.* Minneapolis: University of Minnesota Press, 1960.

Deutsche Welle. "German Cannibal Trial Opens in Complexity." Bonn, March 12, 2003. http://www.dw-world.de/dw/article/0,,1048363,00.html (accessed August 14, 2007).

_____. "German Cannibal Victim's Will Found." Bonn, December 14, 2002. http://www.dw-world.de/dw.article/0,,714339,00.html (accessed August 14, 2007).

Goldman, Lea. "Goriest Show on Earth." *Forbes,* January 30, 2006, 69–71. http://web.ebscohost.com (accessed August 6, 2007).

Harding, Luke. "Victim of Cannibal Agreed to Be Eaten." *Guardian,* December 4, 2003. http://www.guardian.co.uk/germany/article/0,2763,1099477,00.html (accessed August 14, 2007).

Hirschauer, Stefan. "Animated Corpses: Communicating with Post Mortals in an Anatomical Exhibition." *Body & Society* 12, no. 4 (2006): 25–52. http://www.sagepublications.com (accessed August 7, 2007).

Institute for Plastination. *Body Donation for Plastination: Consent Form.* Heidelberg: 2006. http://www.koerperwelten.de/Downloads/BD_Consent_210507.pdf (accessed August 6, 2007).

_____. "Original and Copycat." Gunther von Hagens' Body Worlds: The Original Exhibition of Real Human Bodies. Heidelberg: 2007. http://www.bodyworlds.com/en/exhibitions/original_copycat.html (accessed August 6, 2007).

James, Henry. *The Turn of the Screw.* Edited by Peter G. Beidler. Boston: Bedford/St. Martin's, 1995.

Kuppers, Petra. "Visions of Anatomy: Exhibitions and Dense Bodies." *differences* 15, no. 3 (2004): 123–56.

Landler, Mark. "Eating People Is Wrong! But Is It Homicide? Court to Rule." *New York Times,* December 26, 2003, A4. http://proquest.umi.com (accessed August 14, 2007).

Naughton, Philippe. "German Cannibal Jailed for Life after Retrial." *Times Online,* May 9, 2006. http://www.timesonline.co.uk/tol/news/world/europe/article714862.ece (accessed August 14, 2007).

Pollard, Tom. "Our Lives, Our Bodies: Thorny Issues of Consent." *British Journal of Community Nursing* 7, no. 4 (2002): 172.

Shelley, Mary. *Frankenstein.* New York: Dover, 1994.

Stern, Megan. "Dystopian Anxieties Versus Utopian Ideals: Medicine from *Frankenstein* to *The Visible Human Project* and *Body Worlds.*" *Science as Culture* 15, no. 1 (2006): 61–84.

Stevenson, Robert Louis. *The Strange Case of Dr. Jekyll and Mr. Hyde.* New York: Dover, 1991.

Stoker, Bram. *Dracula.* New York: Oxford University Press, 1983.

von Hagens, Gunther. "Anatomy and Plastination." In Whalley, 9–37.

_____. "On Gruesome Corpses, Gestalt Plastinates and Mandatory Interment." In Whalley, 260–82.

_____. "Why I Think It Is High Time We Democratised the Science of Anatomy." *Times Higher Education Supplement,* March 22, 2002, 18. http://web.lexis-nexis.com (accessed August 7, 2007).

_____ and Angelina Whalley. *Donating Your Body for Plastination,* 7th ed. Heidelburg: Institute for Plastination, 2004. http://slsc.org/uploadedFiles/What_To_Do/BW3Donation-Information(1).pdf (accessed August 6, 2007).

Walter, Tony. "Body Worlds: Clinical Detachment and Anatomical Awe." *Sociology of Health & Illness* 26, no. 4 (2004): 464–88.

Wetz, Franz Josef. "The Dignity of the Individual." In Whalley, 241–59.

Whalley, Angelina, ed. *Gunther von Hagens' Body Worlds: The Anatomical Exhibition of*

Real Human Bodies. Translated by Francis Kelly. Heidelberg: Institute for Plastination, 2006.

Wisnewski, J. Jeremy. "Murder, Cannibalism, and Indirect Suicide: A Philosophical Study of a Recent Case." *Philosophy in the Contemporary World* 14, no. 1 (2007): 11–21.

Individualist Etiologies: Environmental Health, Biological Risk, and Medical Display

Rebecca Onion

At the Body Worlds 1 exhibit, as mounted at the Dallas Museum of Nature and Science in the spring of 2007, a large plexiglass box, filled with packs of cigarettes, matches, and lighters, stands in stark contrast to the scientific tone carefully maintained by the rest of the show. This box is endowed with a placard: "Ditch your pack here and say I Quit!" Nearby, the visitor finds the exhibit's only video display, featuring actor Yul Brynner, who tells the story of his experience with lung cancer, and entreats the viewer never to start smoking.

The bulk of Body Worlds' language, apart from this display, distances the viewer from the processes that led to the death of the bodies on display, withholding information such as age, locale, and, most importantly, name of the deceased. However, this particular section of the exhibit, located with the plastinations intended to showcase cardiovascular systems, reaches out directly to the viewer, entreating her to stop smoking before her lungs turn into black lumps like the ones on display — and, moreover, to take a cathartic public action by dumping the incriminating cigarettes in her pocket right into that plexiglass box.

The only other section of this particular Body Worlds show with a similar agenda of blame and reform is the one centered on cross-sections of a man who had been obese at his death. The intent here is to show how his excess fat, as the exhibit puts it, had "crushed his internal organs," and the language, like that of the smoking exhibit, is exhortative. Although no box is present to receive viewers' stowaway Twinkies and Capri-Suns, the implication is the same: fix your ways, or your own blood may stain your hands.

The Body Worlds catalog echoes these twin fixations. Angelina Whalley, who wrote the bulk of the essay that forms the center of the catalog and covers basic anatomical processes, breaks from very medicalized language to write, in the section on smoking, "Smoking is suicide on the installment plan. As everyone knows, the reason for this is that cigarette smoke can be hazardous to the health and can shorten a smoker's life." This language stands in contrast to the generally clinical tone of the rest of the essay, and six pages of text and pictures are devoted to smoking (as is an entire plastinated body, *The Smoker*). Another rhetorical break comes when the text describes the effects of obesity ("This plastinate shockingly reveals the burden that the inner organs endured during this person's shortened life..."). No such breaks in language occur when the text deals with other sicknesses: arterial sclerosis, melanoma, or fetal malformations.[1]

An unpacking of the Body Worlds exhibit's strange emphasis on smoking and obesity leads to more than a revelation about the exhibit's sometimes-uneven design. Gunther von Hagens and his cohorts are fixated on smoking, to the exemption of other disease-causing factors, because of smoking's capacity to fall into the instructional logic of lifestyle theories of disease. Through an examination of Body Worlds' understanding of how people get sick, we can see how our current model of public health education, as amplified by the showiness of Body Worlds' plastinates, is insufficient for an integrated understanding of human health—an understanding that would incorporate and advance an environmentalist ethic.

Environmental Justice, Environmental Illness, and the Terms of the Debate

Body Worlds' fixation on the individual's responsibility for her own health articulates a dominant cultural model of disease causation. Alternatives to this model, which seek to re-integrate understandings of the environmental causes of sickness, and to find the structural underpinnings of illness, can be found within the environmental justice movement and the medical-historical scholarship which accompanies it. In order to provide a basic understanding as background for the environmental justice arguments I will advance in the remainder of the essay, I will now outline the history and some of the tenets of the environmental justice movement, before returning to an explanation of the Body Worlds etiological worldview.

Although the strain of activism known as the "environmental justice" movement is commonly described as having been born with the publication of Rachel Carson's pesticide protest *Silent Spring* in 1962, historians such as Robert Gottlieb have pointed out that the rise of the idea of the radical connection between the environment and health—and the concurrent movement that would advocate the rights of all people to a healthy environment in which to live—can be traced back to Progressive-era activists seeking to improve conditions for industrial workers.[2] This work formed an important precursor both for Carson and for later activists and groups such as the United Farm Workers, who mounted an anti-pesticide campaign in 1965–1971, and Lois Gibbs, the mother and housewife who spearheaded the anti-toxic-waste movement at Love Canal in 1978. More recently, environmental justice principles have been used to rally citizens in numerous towns in efforts to keep solid waste incinerators, toxic waste dumping, and nuclear plants out of low-income or minority neighborhoods, where they have traditionally been sited.[3]

Environmental justice takes as its central tenet the interconnectedness of humans— especially human bodies, such as the ones exhibited at Body Worlds— with the environment in which they live, a view which "produces new forms of environmental theory and action" along the way.[4] Recently, medical historians, medical geographers, and historians of science have begun to use principles advanced by the environmental justice movement to reconnect histories of health and medicine with histories of the environment. In such books as historian Linda Nash's *Inescapable Ecologies: A History of Environment, Disease, and Knowledge*, a history of human health in California's Central Valley from the time of Anglo-American colonization through the present day, scholars have constructed historical awareness about the interconnections of climate, place, and health.[5] Activists see this re-connection as a potent way to activate environmental concern among citizens, who might otherwise view environmentalism as a highly elite, selective set of concerns.

Body Worlds and the Ethic of Self-Care

Keeping the tenets of environmental justice in mind, the particularly upper class, Western nature of the Body Worlds cosmology of self-care jumps into sharp relief. The individually centered explanation for disease assumes that the individual has a certain amount of control over the substances that enter and exit her body. The rhetoric of blame embedded in the Body Worlds fixation on cigarette smoking and obesity implies that the exhibit viewer could change her ways. Indeed, this rhetoric might almost have the viewer believe that if you are "good" enough, you could live forever — a belief reinforced by the exhibit's dearth of extensive "death stories" of any other, non-smoking or obesity-related type.

The exhibit planners' investment in this individualistic "story" of disease causality can also been seen in official Body Worlds literature. These writings justify the exhibit's very existence through the "evidence" that visitors vowed new attention to their individual bodies after the transformative experience of viewing plastinates. A follow-up survey of visitors in Vienna, Angelina Whalley writes in the Body Worlds catalog, found that "33 percent had been eating healthier food since attending the exhibition; 25 percent had been getting more exercise; 14 percent had generally been living with more awareness of their bodies; and 9 percent had been smoking less since then or had stopped smoking altogether and had been consuming less alcohol."[6] The postcards offered for sale in the Dallas museum gift shop seemed designed as takeaway reminders of the lessons presumably learned in the exhibit about personal care and health. The visitor was able to purchase a card with the smoker's tar-stained lungs on it, as well as one commemorating the plastinated slice of the obese man, juxtaposing it with a coronal slice taken at the same latitude of a man of normal body weight. The legend underneath the two slices reads, "The older you get, the tougher it is to lose weight because by then, your body and your fat are really good friends."[7] The dual chastisement present in this piece of wisdom implies that the person in question is morally deficient, not only for keeping excess weight, but also for growing older at the same time. The card seemed designed to be posted on a refrigerator, a more gruesome version of the common Western cultural practice of affixing pictures of fashion models to the fridge for "inspiration" when faced with the temptation of forbidden foods.

This prescriptive ambition was reiterated, at least at the exhibit in Dallas, by the presence at the exit of a series of pamphlets advertising local businesses that could help the visitor, who would presumably now be on a renewed quest for a type of American-style consumer-oriented fitness. Provided here were two pamphlets for the American Cancer Society, one which gave a list of five potential actions that the reader could take that would prove "lifesaving"—all of them personal, individual, and seemingly simple ("avoid tobacco use; eat right for life—with five or more servings of vegetables and fruits every day; get moving! Be physically active most days of the week; talk to a doctor about cancer screening tests that could save your life; understand your personal risk for cancer and what you can do about it").[8] Alongside these two pamphlets were coupons for a café providing supplements and smoothies, those ubiquitous and expensive concoctions of early twenty-first century corporate fitness culture, and for a complimentary class at a yoga/Pilates/dance studio in the neighborhood.

All of these rhetorical moves for the refinement of individual behavior participate in what environmental justice knows as the *fiction* that humans can control their health if they choose to do so—and that individual rectitude will inevitably lead to health. As political scientist Sylvia Noble Tesh writes, this idea has a history that is intertwined with the basic political assumptions of the American and European worldviews in the nineteenth and twentieth centuries. In the nineteenth century, health crusaders such as Samuel Thomson, who sold botanical secrets under the slogan "Every man his own doctor," and Sylvester Graham, who asked his followers to go vegetarian and eat a low-sugar, low-fat diet, based their appeals on what Tesh calls a "personal behavior theory" of etiology. "In many respects [this theory of disease causation] was a homage to middle-class life," Tesh writes. "It extolled temperate behavior, controlled emotions, minimal physical labor, and the possession of enough money to allow for economic choices."[9] In the twentieth century, this emphasis on "personal behavior" has been officially sanctioned by governmental bodies such as the United States Department of Health and Human Services, which has emphasized what Tesh calls the "lifestyle theory"—a belief that actions that will keep you healthy (basically the same actions recommended by the American Cancer Society pamphlet cited above) demand personal behavior change. Moreover, Tesh points out, adhering to the requirements of the "lifestyle" version of disease prevention serves as a status indicator, staking one's claim on the personality traits associated with the ability to attain longevity—rationality, self-control, foresightedness, and the ability to deny oneself detrimental pleasures.

The lifestyle theory of disease, as embodied by Body Worlds' health prescriptions, places blame on classes of people who are perceived as the opposite of all of the above "positive" traits—people who are, presumably, self-indulgent, lazy, and uninformed. In short, the "unhealthy," under this worldview, embody the traits that are normally used to code the poor or underclass of rich societies.

The environmental justice movement focuses on bringing to light the instances in which human bodies—often, impoverished human bodies—are compromised and sickened by forces beyond the control of the affected persons. For example, many food activists have commented on the way in which coverage of the obesity crisis in America has tended to ignore the fact of unequal access to the kind of food that allows individuals to avoid obesity. Indeed, nutritional researchers have found that the price differential between low-priced, calorie-laden snack foods and more expensive, lower-calorie fruits, whole grains, and vegetables is a major contributor to the obesity crisis among the poor in the United States.[10] When it comes to smoking, the United States Centers for Disease Control reported in 2002 that the link between education and smoking cessation was a strong one—those adults who had attained graduate degrees reported a 7.2 percent smoking rate, while those who had attained General Education Development diplomas (GEDs) reported a 42.3 percent smoking rate.[11] Given correlations between education and income level, it becomes clear that poverty and smoking have a strong relationship. The CDC report posits that this link could be caused by poorer citizens' lack of access to preventive health care and smoking cessation programs. Without stripping smokers and the obese of what conservative political groups would term "personal responsibility," it does seem clear that the story of their illnesses is more complex, less individual, and more tied up with a disadvantaged position in society than the scolding Body Worlds' displays convey.

Body Worlds and the Risk Society: Re-arming the Fiction of Individual Action

Structural problems such as lack of economic access to healthful food and health care could be one component of a citizen's loss of control over her own health, but the even more profound loss of agency comes from environmental degradation. This type of health disruption poses an existential threat, revealing the vulnerability of human life in the face of disturbed biological processes. Analysis of this existential threat will constitute the bulk of the remainder of my essay, since it is precisely this profound realization that Body Worlds cannot at all encompass within its etiological structures.

Here, my analysis begins to follow the work of German sociologist Ulrich Beck. Beck describes the basic tenets of what he calls the "Risk Society," a new, post-industrial order in which the game of accumulating wealth has evolved into the game of avoiding the risks posed by the byproducts of wealth accumulation. Many of these "risks," Beck argues, occur at the most basic level: that of human health. Speaking particularly of radioactivity, pollutions, and pesticides lodged in the groundwater or in the lifecycles of animals, Beck writes: "In developed civilization, which had set out to remove ascriptions, to evolve privacy, and to free people from the constraints of nature and tradition, there is thus emerging a new global ascription of risks, against which individual decisions hardly exist for the simple reason that the toxins and pollutants are interwoven with the natural basis and the elementary life processes of the industrial world."[12] Beck accepts as inevitable the permeability of the body in a post-industrial world. His claim echoes in a more fatalistic way the political protestations of the environmental justice movement, which acknowledges these risks but believes that political action could solve the problems of toxicity with which people are faced.

Those who have more resources, Beck writes, will be able to avoid some of these health risks through following the demonstrated model of the lifestyle theory of disease: "'A sufficiently well filled wallet puts one in a position to dine on eggs from 'contented hens' and salads from 'pampered heads of lettuce.' Education and attentiveness to information open up new possibilities of dealing with and avoiding risks."[13] Thus, for example, children in inner cities suffer from asthma at a high rate, and the nation of China, full of a very politically vulnerable populace, experiences environmental illness on such a level that pollution-caused cancers are the number one cause of death.[14]

However, Beck argues, even those phenomenonally well-endowed with resources, both financial and political, will not be able to escape the risks completely. This principle can be seen on a macro level in the example of the overwhelming dangers presented to long-term physical well-being by global warming, and on a micro level in cases of cancer caused by sun exposure or by pollution in urban areas. Body Worlds' participation in the particular individualistic story of disease marks it as a product of a structure of knowledge that is inadequate for the physical reality of the embodied, post-industrial biosphere.

Commentators have examined the relationship of Body Worlds to art, pointing out how several of the plastinates echo famous paintings or sculptures (an example is Jose van Dijck's comparison of *The Chess-Player* to Rodin's *The Thinker*, or of the plastinate *The Runner*, whose muscles flutter behind him in the wind, to the kinetic work of futurist art).[15] But a more apt comparison may be to the example of the classic natural history museum of the early twentieth century, in which animals and humans of a "lower" order are shown in their "habitats," surrounded by the other animals and plants which form their bioworld. Thus, the Giant of Karisimbi, the prized silverback gorilla of the American Museum of Natural History, which was shot by the collector Carl Akeley, is shown in a diorama with other members of its social group. The gorillas are placed in a forest setting, with the mountains of its habitat visible in the background. The Museum of Natural History, as Donna Haraway has written, also displayed (and currently displays) non–European people as components of this unbroken "natural history," implicitly endowing the (white) American viewers of the museum with a higher, post-"natural" status.[16] These exhibits all emphasize the contingent existence of the animal or the Native American or the Samoan, framing each as a being dependent on the biological substances of its habitat, and subject to observation by the presumed Euro-American museum visitor.

In contrast to these natural-historical situational exhibits or displays, Body Worlds plastinates are endowed with positions that emphasize their continuing humanity, even in the non- or anti-human state of death. The actions they perform are non-essential for survival — trademark "human" behaviors: playing soccer, playing chess, teaching in a classroom setting, skateboarding, and riding a horse. In an essay in the exhibit's catalog on the process of choosing the poses of the plastinates, von Hagens writes that these poses were chosen in order to facilitate connection with the bodies: "One of the essential design principles of the Body Worlds exhibition has been based on the conviction that a lifelike and attractive setting, e.g. one enlivened with plants and objects from daily life, would facilitate dealing with human specimens..." His examples include the props in the plastinates *The Chess Player* and *The Swordsman*, but he also writes, "Other gestalt plastinates have been placed in a garden setting complete with a brook."[17] The description implies that the landscape into which the plastinates have been put is a "cultured" nature. That is, the landscape itself would also show evidence of human cultivation, reinforcing the dominant story of the gestalt plastinates: this human body stands apart from the messy or uncontrolled life of an animal, establishing itself as one of a member of a species which actively shapes its environment.

This story, which reinforces the individualistic etiology of bodily disorder described above, emphasizes the traits that define the "human." In this view, the "human" is an entity which has the ability to perform actions, be they directed toward leisure, transport, self-care, or self-improvement, that cannot be realized by the masses of animals who share our basic physiological makeup. In this context, vulnerability to disease and death, which is a basic condition of the embodied life, is denied by the ongoing ability of these plastinates to perform "human" actions. As van Dijck points out, von Hagens' continuing insistence of the "authenticity" of these bodies is beside the point: "The modified nature [of the plastinates] is the very reason why his plastinated corpses can be patented... Although in many ways his work is a continuation of age-old traditions in the material production of anatomical objects, he also adds a commentary on the 'nature' of the human body: humans are no longer subjected to divine nature, as science and scientists to a large extent control longevity and quality of life."[18] For van Dijck, von Hagens' plastinates are proof of the "posthumanity" of the human body, as exemplified by the plastinate named *The Orthopedic Body*, which underwent a large variety of postmortem surgeries, which have left it with many metallic appendages and joints. These plastinates display in and of themselves the great power of human artifice to transcend the grip of natural laws.

Thus, the interesting distinction to be made, for the viewer concerned with re-placing the human body into the environment, is not between the Body Worlds plastinates and traditional art objects, but rather between the Body Worlds plastinates and objects in a natural history museum. In this distinction lies the inherent human fiction of our difference from the animal, which supposedly lies in our ability to control nature by "keeping" ourselves well; results in our evolving complicated extraneous "human"-branded actions such as the playing of soccer or chess; and allows for the very existence of the Body Worlds phenomenon. A quote from Immanuel Kant, printed on a large banner and used as background to the exhibit, holds some kind of a key to this ongoing distinguishing between man (the self-contained unit) and animal (the affected, the powerless). The quote reads: "That man can be conscious of himself in his contemplation raises him infinitely above all other living creatures of earth."

The environmental perspective on these fictions lies in the need to break down these boundaries between the human and the non-human. Realization of the vulnerability of human flesh can, of course, be a profoundly disturbing experience, and this is why news of the dangers of radioactivity, pesticides, and pollution has such power to bring home to us our own permeability. The logic of Body Worlds would separate and insulate the human from the vulnerable animal, marking the human as the "animal," which has the power of self-contemplation, healing, and immortality. Within this cosmology, news of radical threat from the biological entities around us— entities that we ourselves have, ironically, created or fouled —can find little purchase. It is for this reason that the concerns of environmental health and environmental justice face an uphill battle in their fight for recognition. The Western tradition, as self-consciously epitomized in Body Worlds, is fundamentally exclusionary to its realization.

Conclusion: Body Worlds and the (non)Creation of Risk Knowledge

> [These risks] induce systematic and often irreversible harm, generally remain invisible, are based on causal interpretations, and thus initially only exist in terms of the (scientific or anti-scientific) knowledge about them. They can thus be changed, magnified, dramatized, or minimized within knowledge, and to that extent they are particularly open to social definition and construction. Hence the mass media and the scientific and legal professions in charge of defining risks become key social and political positions.[19]

In this passage, Ulrich Beck points out that the types of risks faced by bodies in a post-industrial world are not easily categorizable by our current structures of knowledge. It is for this reason that cancer clusters caused by pollution, for example, are not often officially "verified": epidemiological structures have evolved to require definitive proof of causation. When correlation is all that can be found, sufferers find it difficult to get confirmation of their sicknesses' relationship to environmental causes.[20] Similarly, when sufferers of multiple chemical sensitivity attempt to have their sickness medically confirmed, they face barriers partially erected by structures of medical knowledge.[21]

The example of Body Worlds shows how these structures of knowledge also control public awareness of disease etiology, creating a dominant story about disease in which humans, isolated from their environments, either remain healthy by dint of their own efforts to do so, or fall ill due to their moral failings. Interestingly, Body Worlds sees public education about the workings of the body as part of its mission. In a forward to the exhibit's catalog, anatomist Wilhelm Kriz, a former professor of von Hagens, writes, "Medical laypersons too have the right to ask questions about how the human body looks on the inside, and they too have a right to want to see it. They have a right to learn where and how a disease occurs and how it might be prevented if possible."[22] In the throes of the rhetoric of the importance of scientific truth, the exhibit, however, omits and eliminates versions of disease stories that do not reinforce a narrative of human control on the scientific and individual levels. This dominant narrative is one reason why the efforts of environmental justice activists to reactivate knowledge of our bodily environmental connections are so radical and so potentially revolutionary.

Notes

1. Whalley, "The Human Body," 68, 190–91, 84–5, 97,127.
2. Gottlieb, *Forcing the Spring.*
3. Bullard, *Dumping in Dixie*; Di Chiro, "Nature as Community," 298–320; Szasz, *Ecopopulism*; Pulido, *Environmentalism.*
4. Di Chiro, 300.
5. Nash, *Inescapable Ecologies.* See also Morris, "Environment: The White Noise of Health"; Mitman, Murphy, and Sellers, "A Cloud Over History"; Kearns and Geisler, "Introduction."
6. Whalley, "Body Worlds," 303.
7. Institute for Plastination, "Obesity Revealed."
8. American Cancer Society, "Five Lifesaving Things."
9. Tesh, *Hidden Arguments,* 24.
10. University of Washington, "Researcher Links."
11. Centers for Disease Control, "Cigarette Smoking."
12. Beck, *Risk Society,* 41.
13. Ibid., 35.
14. Kahn and Yardley, "As China Roars."
15. van Dijck, "Bodyworlds," 99–126.
16. Haraway, "Teddy Bear Patriarchy," passim.
17. von Hagens, "On Gruesome Corpses," 266.
18. van Dijck, 111.
19. Beck, 22–23.
20. See Nash, chapter 5, "Contesting the Space of Disease."
21. See Kroll-Smith and Floyd, *Bodies in Protest.*
22. Kriz, "Foreword," 7.

Bibliography

American Cancer Society. *Five Lifesaving Things You Can Do.* Oklahoma City, 2002.

Beck, Ulrich. *Risk Society: Towards a New Modernity.* London: Sage Publications, 1992.

Bullard, Robert. *Dumping in Dixie: Race, Class, and Environmental Quality.* Boulder, CO: Westview Press, 2000.

Centers for Disease Control. "Cigarette Smoking Among Adults— United States, 2002." Centers for Disease Control and Prevention. May 28, 2004. http://www.cdc.gov/mmwr/preview/mmwrhtml/mm5320a2.htm. (accessed January 19, 2008).

Di Chiro, Giovanna. "Nature as Community: The Convergence of Environment and Social Justice." In *Uncommon Ground: Rethinking the Human Place in Nature,* edited by William Cronon. New York: Norton, 1996. 298–320.

Gottlieb, Robert. *Forcing the Spring: The Transformation of the American Environmental Movement.* Washington, DC: Island Press, 1993.

Haraway, Donna. "Teddy Bear Patriarchy: Taxidermy in the Garden of Eden, New York City, 1908–1936." *Social Text* 11 (1984–1985): 20–64.

Institute for Plastination. "Obesity Revealed." Postcard. Gunther von Hagens' Body Worlds: The Original Exhibition of Real Human Bodies. Heidelberg: 2007.

Kahn, Joseph, and Jim Yardley. "As China Roars, Pollution Reaches Deadly Extremes." The *New York Times,* August 25, 2007. http://www.nytimes.com/2007/08/26/world/asia/26china.html (accessed January 19, 2008).

Kearns, Robin A., and Wilbert M. Gesler, "Introduction." In *Putting Health in Place: Landscape, Identity and Well-Being,* edited by Robin A. Kearns and Wilbert M. Gesler. Syracuse, NY: Syracuse University Press, 1998. 1–16.

Kriz, Wilhelm. "Foreword." In Whalley, *Gunther von Hagens',* 5–8.

Kroll-Smith, Steve, and H. Hugh Floyd. *Bodies in Protest: Environmental Illness and the Struggle Over Medical Knowledge.* New York: New York University Press, 1997.

Mitman, Gregg, Michelle Murphy, and Christopher Sellers. "A Cloud Over History." *Osiris* 19 (2004): 1–17.

Morris, David. "Environment: The White Noise of Health." *Literature and Medicine* 15, no. 1 (1996): 1–15.

Nash, Linda. *Inescapable Ecologies: A History of Environment, Disease, and Knowledge*. Berkeley: University of California Press, 2006.

Pulido, Laura. *Environmentalism and Economic Justice: Two Chicano Struggles in the Southwest*. Tucson: University of Arizona Press, 1996.

Szasz, Andrew. *Ecopopulism: Toxic Waste and the Movement for Environmental Justice*. Minneapolis: University of Minnesota Press, 1994.

Tesh, Sylvia Noble. *Hidden Arguments: Political Ideology and Disease Prevention Policy*. New Brunswick: Rutgers University Press, 1988.

University of Washington. "Researcher Links Rising Tide of Obesity to Food Prices." *ScienceDaily*, January 5, 2004. http://www.sciencedaily.com/releases/2004/01/040105071229.htm (accessed January 19, 2008).

van Dijck, Jose. "Bodyworlds: The Art of Plastinated Cadavers." *Configurations* 9 (2001): 99–126.

von Hagens, Gunther. "On Gruesome Corpses, Gestalt Plastinates and Mandatory Interment." In Whalley, 260–282.

Whalley, Angelina. "Body Worlds Through the Eyes of Its Visitors." In Whalley, 299–303.

_____. *Gunther von Hagens' Body Worlds: The Anatomical Exhibition of Real Human Bodies*. Translated by Francis Kelly. Heidelberg: Arts and Sciences, 2006.

_____. "The Human Body—Anatomy and Function." In Whalley, 38–207.

Section Two

The Usual Gruesome Anatomical Apparitions

The Persistence of Tradition in Anatomical Museums

Stephen Johnson

... no one who enters this place of anatomical science, from the medical student to the uninformed visitor, will leave it without having added something to his previous store of knowledge, which (as it is useful) may be of use to him. And it is to be hoped, that, by seeing the various organs and functions of the human frame, with their varied uses and operations, it will be recollected what miseries may be entailed by a self-willed and wanton destruction of the body... The proprietor has the satisfaction of knowing that his endeavours have not been unavailing. Many a young person has been saved from vice by an inspection of his Museum.
—from the catalog for Jacob Reimers' Anatomical Museum, 1853[1]

Laypersons have retained a fine sense for authenticity. They have a craving for it... Aestheticism and dynamics foster instruction.
— Gunther von Hagens [2]

The range of visitor reactions to the Body Worlds exhibitions of the past ten years has been disparate, contradictory, and complex. Medical and religious organizations and citizens' groups have attempted to shut them down, while crowds have lined up for hours to enter. Some visitors faint, requiring medical attention and an early exit, while parents with small children blithely discuss every specimen. Gunther von Hagens, the creator of these exhibitions and the process of plastination—by which the fluids of a human corpse are drained and replaced by polymers, resulting in a preserved specimen of tissue previously impossible—has been by turns vilified as a new Frankenstein, a grave robber and a charlatan, and praised as a great educator.[3] The Body Worlds exhibitions, despite or because of their ongoing controversy, have been visited by millions on three continents, making them an important site for the study of the ways in which people look at the body.

This essay focuses specifically on the persistence and erasure of traditions of bodily display in Body Worlds. It first of all details and assesses my own attendance at the exhibition in the autumn of 2002, with the objective of "getting to" some of the complexity of a visitor's reaction at this early stage in its brief history. It then examines that experience of Body Worlds as compared with one of its historical antecedents, the Victorian anatomical museum. Looking at catalogs and posters from the London-based institutions of J. W. Reimers and Joseph Kahn, a number of similarities emerge with Body Worlds which suggest that, in some measure, what it is that unsettles (some of) its visitors is a disruptive image of past practice. With this in mind, the essay ends with another report, this time of my visit to Body Worlds 2, in which Victorian historical tradition has been further excised, making the exhibition more palatable for a broader audience. In each case the discussion betrays the prejudice that as an historian, I cannot experience any contemporary phenomenon without looking for the remnants of the past; and I cannot appreciate the present without seeing both the persistence of tradition embedded in what I am experiencing and its adaptation to changing circumstance. The longevity of a tradition may be obscured, if not erased, by recent innovation, but often the past rests buried under the accumulated detritus of change.

Body Worlds, Old Truman Brewery, Brick Lane, London, Autumn 2002

At this point in the brief history of the Body Worlds exhibitions von Hagens, though successful, still had some difficulty gaining institutional respectability; for this reason, the London version of Body Worlds was located in an unlikely neighborhood in the east end of London, in a disused factory. Visitors entered down a long hallway that displayed hundreds of press clippings tracing the development of the process of plastination, the display of its results, and in particular all of its controversies—questions about the source of the bodies, the ethics of display, the respect or disrespect to the body, and so on.* It was hardly possible to enter the exhibition without a knowledge of, and a focus on, the controversy.

The exhibition itself was set in a large, high-ceilinged, garden-like setting, the regularly-placed foliage interspersed with four kinds of exhibit. In an array of glass cases plastinated specimens were displayed according to function and pathology—the respiratory, digestive system, and reproductive systems, for

**Partial list: "Ghoulish? Our kids,"* Sunday Express; *"Artist or Butcher?,"* The Observer; *"Lecturer Takes Hammer,"* The Times; *"Vile Bodies,"* The Guardian; *"The Grave Robber?,"* Mail on Sunday; *"Is Sucking the Flesh,"* Globe and Mail; *"'Dr Frankenstein' defends,"* Daily Telegraph; *"A Grim Anatomy Lesson,"* Catholic Herald; *"A Body World of Troubles,"* East London Advertiser; *"Grim Show gets the go-ahead,"* Metro London; *"Is this art?,"* Herald; *"Freak Shows,"* Manchester Evening News; *"Dr Frankenstein, I presume,"* Independent; *"How low will people?,"* Ottawa Citizen; *"Show of dissected corpses,"* Guardian.

example. Displays showed healthy and smokers' lungs, or healthy brains with examples of stroke patients. The second kind of display was the translucent "sagittal," consisting of horizontally or vertically sliced plastinates arranged to show the relationship of a whole body to its parts. The third kind of exhibit was the "configuration," created by preserving the circulatory system of a whole or part of a body, along with its structure, and then eliminating the rest of the body. The results were recognizable as people and animals, but comprised entirely of red-dyed blood vessels. These were all impressive, of course, but the basis for the overall structure of the exhibition, and the reason for its popularity, was the presence of whole body plastinates, for which the entire corpse had been flayed, manipulated to some effect, and posed. Many of these were posed naturally: playing chess or basketball, pole-vaulting, swimming, fencing, and so on. Some were manipulated beyond the "natural" pose to show off parts of the anatomy: *The Pole-Vaulter* was frozen in mid-leap upside-down, his inner organs removed and placed nearby, also upside down; *The Fencer* was split vertically into three parts set side by side; and *The Swimmer* was split horizontally into two and set hanging in mid-air. Visitors clustered around these plastinates throughout the exhibition, examining them closely, reading the signage, listening to their audiotapes, talking.

Although there was a generally visible path through the hall, visitors could wander. The central orienting figure was a large rearing horse and rider; both were flayed and plastinated, the body of the rider split vertically into three parts. The exhibition became, in my opinion, more diffuse, and unusual, as I proceeded toward the exit. Whole body plastinates were replaced by the circulatory models, sagittals were arrayed in cases that were not specific to a function or lesson, and an alcove to one side concentrated on the development of the fetus (to be discussed later). The end of the exhibition presented information on the process of plastination, prior to a single exit into the gift-shop.

In retrospect, I can attribute three general characteristics to this exhibition:

1. Reference to a long tradition. Large signs reprinted imagery from the Renaissance, showing posed nudes as well as anatomical specimens; some whole body plastinates imitated the originals specifically and in proximity. Written literature, both accompanying the specimens and printed in the catalogs, repeated this claim, with an emphasis placed on passing down the skill of the artist and the scientific knowledge of the anatomist.[4]
2. A pedagogical imperative. The signage on all specimens, the voice on the accompanying audio-tour, the large banners in the hallways, all promoted the journey of the visitor as a means to learn about the body. Positions of bodies, however obscure in some cases, were explained as best suited to showing off a function (the nervous system, for example). As noted, specimens

were typically paired to show the effects of disease, or environmental hazards. In general, the entire display was presented — and defended — as an aid to a democratic understanding of the body.

3. An aesthetic insistence. Just as certainly as the exhibition promoted pedagogy, it also promoted the creative artistry of the anatomist in the plastination of the body. Although not perhaps as insistently stated in the written context to the specimens, the audiotape repeatedly emphasized that the "aesthetic position is intentional" and that the process is "an art as well as a science."

These general characteristics are consistent with the means and ends of Body Worlds as von Hagens states them. He believes that audiences have a "craving" for authenticity, and the process of plastination fulfills that longing.[5] At the same time, he believes that the means to persuade spectators to consider their own bodies in greater detail, to think about how they are used and abused, and to enter into what he believes is a partnership with the medical profession, is through an aestheticization of the body provided by the anatomist (see opening quotation).[6] He believes that the manipulation of the body in ways that are aesthetically pleasing or arresting — often mimicking the traditions that have associated artistic expression and anatomical study — advocate for an interest in and a respect for the body. It may, as well, become more acceptable to view. This is reinforced in a DVD interview:

> Laypersons have a right to view the human body, so that they can form an impression of its anatomy, so that they can genuinely comprehend the body... This exhibition has been consciously made for laypersons. The reason is that laypersons are the partners of today's physicians. Consequently they have to be educated about the human body. They can only receive this education when they look inside genuine human bodies, just as physicians were able to do during their studies.[7]

Venerable tradition, democratic learning, and aesthetic manipulation are all clearly visible contextual strategies to guide audience reception at Body Worlds. During my visit, however, I also observed an uneasy, ambivalent relationship with spectators. Although any eyewitness description of audience response must be treated with suspicion, it seemed to me that the reactions of visitors shifted markedly when confronted by the whole body plastinates. Reaction to the body in its parts was sedate, apparently accepting the pedagogical context; but around the whole body specimens there was jostling, noise, a good deal of wry comment, and a degree of mockery, as if spectators, when confronted by an exposed and manipulated human being to which they must react, do so through dismissal or humor, or both. There were, no doubt, a full range of responses, from the sympathetic through the analytical to the parodic — but, of course, only the last of these "acts out" in any memorable way to a bystander.

There is, I believe, a relationship between this unease and several aspects of this Body Worlds exhibition that I found at odds with the exhibition's general motives:

1. I was struck, from time to time, by the presence of distinct individual characteristics in specimens. In general, the whole body plastinates were, by late-twentieth-century standards of beauty, athletic, well-shaped, flat-bellied, firm-breasted, well-endowed. They were presented as general, ideal representations of the human body, rather than the remains of a particular cadaver. There were no names, no individual traits, and no particular sense that the task to which each specimen had been assigned was something they might have done in life. *The Basketball Player* may never have seen a basketball.

And yet there were disruptive intimations of individuality. I found it frankly unnerving whenever skin was present on a specimen, accompanied by vestiges of hair, and in one case by a tattoo. Similarly, I repeatedly contemplated — and overheard others discussing — the origin of the specimens, many of whom appeared to be of Asian descent (whether this was in fact the case, von Hagens had business interests in Kyrgyzstan and China).[8] The presence of skin and the cultural origin of the body are less important than the fact that, given a remaining clue to individuality exposed beneath the aestheticization (and, apparently, effacement) of the body, the visitor (myself) was arrested by the prospect of a personal history. I had similar reactions to such mundane items as toenails, and to crooked and missing teeth, which to me were at odds with both pedagogical and aesthetic intentions. In a very real way, these individuals had been de-personalized by removing everything we might have seen of them in life, first of all by flaying, and then by creating a new persona. And yet, signs of individuality remained that distracted me from both "pedagogy" and "aestheticism."

2. In a few instances the manipulations of the body by the anatomist appeared to leave the pedagogical imperative behind. Some aberrations were minor: *The Winged Man*, for example, wore a hat on a particularly grotesque face, perhaps intended to add humor or humanity. Other specimens moved further to the edges of the pedagogical imperative. Specifically, one figure was posed kneeling in prayer, holding aloft his heart; behind him was his historical antecedent, a Renaissance print in the same pose (without the heart). This figure was dedicated "to the donors," its theological intention clear; it was, in its effect, an allegorical figure representing all of the donors who were not present as individuals. Like the small vestiges of hair or tattoo, it was a rare reference to the origin of the specimens, foregrounding the general anonymity and raising unintended questions. Were the donors, for example, Christian?

But the most extraordinary manipulation of a body in this version of Body Worlds, and one that, I believe, has not been in evidence since, was *The Mystical Plastinate*. This figure was placed at the end of the exhibition, near a display about the process of plastination and the museum shop, and in a small alcove on its own. This specimen was hanging freely, facing the viewer, his/its head perhaps three feet higher than its feet, at an angle. Oversized glasses sat on its head, magnifying its eyes out of all proportion to the body. Its wrist, ankles, and spine muscles were splayed out in strips from the body. Its face had been peeled up from the skull into a pointed flap above the head, like a fool's cap. It was holding in its hands, as if presented to the spectator, its inner organs. It is not easy to capture the effect of this image in words (nor to determine which gender to use); the explanatory note next to the exhibit makes the attempt:

> Fairy tales with Mystical Creatures belong to our cultural heritage. They create a world of fantasy and humor making our mortality easier to accept. Here you see the first plastinate of this kind. The inner organs from the tongue to the rectum are presented in front of the thoracic and abdominal cavity. The right hand holds the tongue and larynx while the left hand supports the left lung. The spinal cord is exposed by expanding the surrounding back muscles and the body spine. The muscles of the limbs have been flapped back from their fleshy origins, demonstrating their tendoneous [sic] insertions and the long bones below. The facial muscles are presented from the inside by flapping the soft tissue of the face up above the head like a pointed cap. The glasses of the plastinate serve to magnify the eyeball and eye muscles.

For me, this was certainly the most troublesome specimen in the exhibition; and it may be significant that I can find no record of it except in this exhibition. I also could not at the time understand what place it had in the general tenor, tone, and intention of Body Worlds. It invoked a world of myth and of humor. It was, in art-historical terms, a true grotesque, far removed from the well-proportioned classical symmetries of the Renaissance artist-anatomists invoked elsewhere. If traditions and antecedents were at work, this had more of Rabelais than of Michelangelo in it.

3. Finally, I was struck by the singular inclusion of natural deformities. This may not at first seem out of place in an exhibition stressing pedagogy—pathology is a normal part of an anatomical education. Healthy and diseased lungs, brains, and digestive tracts were all displayed to effect. However, there were restricted to one alcove plastinates of aborted fetuses that had suffered from Hydrocephaly, Anencephaly, and Cleft Palate, as well as Conjoined ("Siamese") twins. These were exceptional specimens, treated in an exceptional manner by the visitors around me. Confronted by deformity, there was a notable increase in the already evident jostling, humor, and sarcasm of visitors, as if in increased compensation for the discomfort the displays were causing.

Persistence of Tradition: Reimers and Kahn in the 1850s

The effort to understand a culture is measured by those aspects that fall outside of its apparent norms, and yet persist in the tradition (the "exceptional normal").[9] As surely as the joke we don't understand, or the regular presence of the unacceptable, we can only "get to" the complexity of an event through its incomprehensible elements. I have attempted to outline some aspects of the Body Worlds exhibition I witnessed in London that appeared to me to stand outside of the stated goals of the organizers. Gunther von Hagens clearly prizes historical traditions of anatomical display; it is this interest that helps to explain my own discomfort. Specifically, more than the often-mentioned Renaissance artist, Body Worlds bears so many clear correspondences with the Victorian popular anatomical museum that, in my opinion, the imitation must have been conscious.

A visitor to Kahn's or Reimers' anatomical museums in 1850s London would have been treated to two kinds of exhibition recognizable to the visitors from Body Worlds.* First of all, there were lecture-demonstrations using an *Anatomical Venus* or *Anatomical Adonis*, a whole body wax (or *papier maché*) model that at first glance was a reclining, sleeping nude.[10] A lecturer (typically a man for a male audience, a woman for a separate female audience) would conduct a kind of verbal tour of the body, speaking to the audience as they gathered around, opening up the model to remove a series of body parts while providing instruction. In addition, these museums included enlarged models of eyes, ears, and other parts of the body, as well as accurate representations of the body's various systems, according to the science of the day. A sample "tour" of Dr. Kahn's Anatomical Museum was published in the *Era*, 30 March 1851:

> We see preparations from life, and others of what once belonged to life itself, models in leather, casts in wax, realities so carefully preserved, similitudes so striking, imitations so exquisite, and illustrations so perfect, that the mind is bewildered by the reflections they suggest, and the truths they reveal. Here we have the economy of the human frame laid bare in every particular, and the mysteries of mortality unfolded with artistic skill and minute precision, at once a curious and an edifying sight. We see anatomical figures taken to pieces bit by bit, and the process of digestion explained. We behold large models of the eye and the ear, and strange degrees of comparison. A lecturer attends, and with his subject literally at his fingers' ends, explains, with remarkable ease and perspicuity, all that is before you. You have particular organs and functions cleverly described, structures so finely moulded that you cannot help admiring the ingenuity of man, even while you see the surprising workmanship of his Maker. Arteries, veins, and muscles, are represented plainly and accurately. But a series of figures exhibiting the progress of a human being, from the earliest origin that can be perceived, to maturity is of itself a remarkable collection of figures, and here they are, for the lovers of science and the explorer of nature. We see what tight lacing will do, and what more reprehensible acts will bring to pass—"a sorry sight!" The visitor, however, must not imagine that he goes to explore an exhibition similar to that of Madame Tussaud's. These are works of art—subjects for philosophical contemplation. Having overcome a natural repugnance, you examine

**A range of playbills and catalogs examined for this essay are located in the John Johnson Collection, Bodleian Library, Oxford University; the Wellcome Library for the History of Medicine, London; and the British Library. In particular: Joseph Kahn,* Catalogue of Dr. Kahn's Anatomical Museum *(London, 1851), which lists and describes 341 items; and J. W. Reimers,* J. W. Reimers's Gallery of All Nations and Anatomical Museum *(Leeds, 1853), which lists 410, and claims more than 450 items on display.*

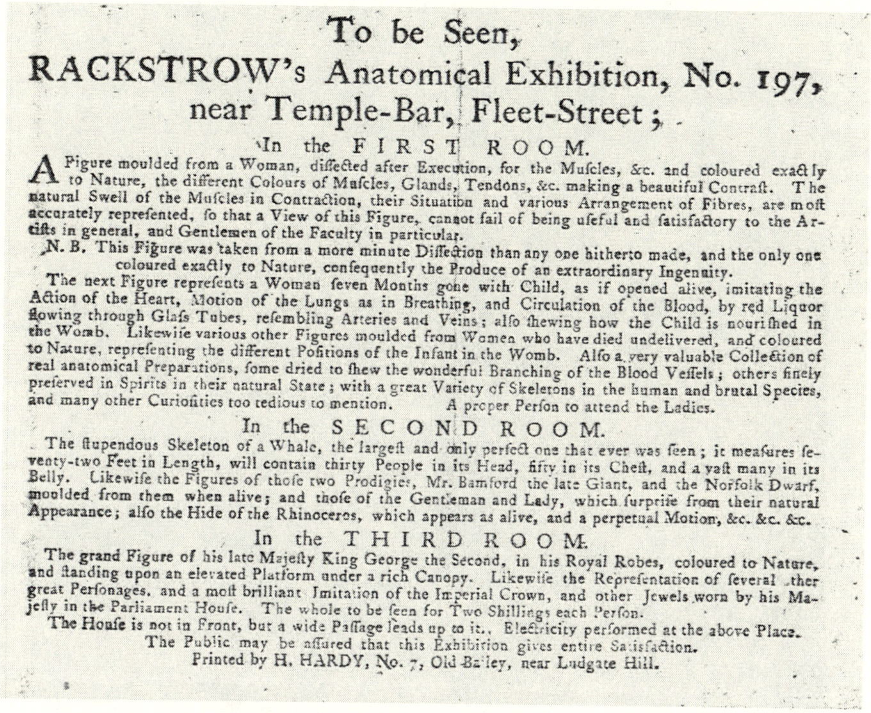

To be Seen,

RACKSTROW's Anatomical Exhibition, No. 197,

near Temple-Bar, Fleet-Street;

In the FIRST ROOM.

A Figure moulded from a Woman, diſſected after Execution, for the Muſcles, &c. and coloured exactly to Nature, the different Colours of Muſcles, Glands, Tendons, &c. making a beautiful Contraſt. The natural Swell of the Muſcles in Contraction, their Situation and various Arrangement of Fibres, are moſt accurately repreſented, ſo that a View of this Figure, cannot fail of being uſeful and ſatisfactory to the Artiſts in general, and Gentlemen of the Faculty in particular.

N. B. This Figure was taken from a more minute Diſſection than any one hitherto made, and the only one coloured exactly to Nature, conſequently the Produce of an extraordinary Ingenuity.

The next Figure repreſents a Woman ſeven Months gone with Child, as if opened alive, imitating the Action of the Heart, Motion of the Lungs as in Breathing, and Circulation of the Blood, by red Liquor flowing through Glaſs Tubes, reſembling Arteries and Veins; alſo ſhewing how the Child is nouriſhed in the Womb. Likewiſe various other Figures moulded from Women who have died undelivered, and coloured to Nature, repreſenting the different Poſitions of the Infant in the Womb. Alſo a very valuable Collection of real anatomical Preparations, ſome dried to ſhew the wonderful Branching of the Blood Veſſels; others finely preſerved in Spirits in their natural State; with a great Variety of Skeletons in the human and brutal Species, and many other Curioſities too tedious to mention. A proper Perſon to attend the Ladies.

In the SECOND ROOM.

The ſtupendous Skeleton of a Whale, the largeſt and only perfect one that ever was ſeen; it meaſures ſeventy-two Feet in Length, will contain thirty People in its Head, fifty in its Cheſt, and a vaſt many in its Belly. Likewiſe the Figures of thoſe two Prodigies, Mr. Bamford the late Giant, and the Norfolk Dwarf, moulded from them when alive; and thoſe of the Gentleman and Lady, which ſurpriſe from their natural Appearance; alſo the Hide of the Rhinoceros, which appears as alive, and a perpetual Motion, &c. &c. &c.

In the THIRD ROOM.

The grand Figure of his late Majeſty King George the Second, in his Royal Robes, coloured to Nature, and ſtanding upon an elevated Platform under a rich Canopy. Likewiſe the Repreſentation of ſeveral other great Perſonages, and a moſt brilliant Imitation of the Imperial Crown, and other Jewels worn by his Majeſty in the Parliament Houſe. The whole to be ſeen for Two Shillings each Perſon.

The Houſe is not in Front, but a wide Paſſage leads up to it. Electricity performed at the above Place.

The Public may be aſſured that this Exhibition gives entire Satisfaction.

Printed by H. HARDY, No. 7, Old Bailey, near Ludgate Hill.

Figure 3: Bill advertising the early appearance of an anatomical museum in London, circa 1782, including figures representing women "Months gone with Child, as if opened alive..." John Johnson Collection, Waxworks 3 (31) and (38), Bodleian Library, University of Oxford.

> what you see for the sake of the knowledge to be gathered, and you come away with mingled humiliation and awe—a profitable impression.[11]

The correspondence to Body Worlds is obvious, the variations primarily technological; the stationary live lecture interspersed among the galleries has been replaced by a mobile audio-tour. Even the twin emphasis on pedagogy and aesthetic value in Body Worlds is reflected in the Victorian museum through the advertised skill of the modeler—just as von Hagens' name is displayed on each plastinate. Similarly, Victorian museums displayed the effects of environment and personal choice on the body, exposing the terrible results of "the tight lacing of corsets" and "onanism." Time has changed the social evils to smoking and obesity, but the function of the exhibition is the same.

This may be true more generally. Historians of the Victorian anatomical movement give it a specific set of cultural goals: the education of the working class by the medical profession, and the teaching of an "anatomical world view." Using Foucault as a starting point, historians Renée Burmeister, Ruth Richard-

Dr. KAHN'S ANATOMICAL MUSEUM,

4, COVENTRY STREET

LEICESTER SQUARE,

OPEN DAILY (FOR GENTLEMEN ONLY)

FROM TEN TILL TEN.

An entirely New Series of highly interesting Lectures

Delivered daily by **Dr. KAHN,** at ½-past **8** p.m.

Also by Dr. SEXTON, F.R.G.S., F.E.S., &c.

At **12, 2,** and **4**; and Half-past **7** in the Evening.

ADMISSION, ONE SHILLING.

☞ NONE BUT ADULTS ARE ADMITTED.

The above Unique Collection contains nearly **ONE THOUSAND** highly-interesting and superbly-executed Models, illustrating every part of the Human Frame, in **HEALTH** and **DISEASE.**

The following are the principal Models and Preparations:

THE DEVELOPMENT of THE HUMAN EMBRYO, from the first impregnation of the egg to the period of birth.

☞ *These facts are shown both by highly magnified Models, and also by Natural Preparations in Spirits, and are the only series of the kind in the world.*

THE FIVE SENSES, highly magnified.

THE APOLLO BELVIDERE, and **THE VENUS DE MEDICI,** figures which take to pieces, and show every part of the human body, and to perfect which, **300 Dissections** were made.

A successful case of **THE CÆSARIAN OPERATION,** the Lady being still living, at Paris.

A full-length Model of a Young Lady who fell dead at a ball, from the pernicious **EFFECTS OF TIGHT LACING.**

Models showing the Operations for **Cataract** and **Squinting,** the **Rhinoplastic Operations** for restoring a lost Nose, and a full-length Model illustrating the **Operation for Lithotrity,** or crushing the Stone in the Bladder.

A collection of **50 Busts** of the various **Races of Men,** the *most perfect Ethnological Series in the world.*

THE VARIOUS OPERATIONS IN MIDWIFERY.

Numerous MALFORMATIONS & MONSTROSITIES.

In addition to the above, and hundreds of others, Dr. KAHN specially calls attention (more particularly of young men) to the **PATHOLOGICAL ROOM,** where the various diseases arising from the errors during the critical period from adolescence to manhood are accurately portrayed, and where may be had—A LESSON FOR ALL!

A Descriptive CATALOGUE, to which is also added several LECTURES as delivered by Dr. KAHN daily, under the title of

"SHOALS AND QUICKSANDS OF YOUTH,"

Will be presented GRATIS to every visitor to the Museum, or may be had by Post, on enclosing 12 Stamps to the Author.

son, and Michael Sappol (among others) argue that the anatomical museum at this time represented a concerted effort to improve the status of the surgical profession.[12] Surgeons were overcoming grave-robbing scandals, fears concerning the institutionalization associated with hospitals, and the general resistance to the idea that one could cure through cutting. Teaching citizens about the body's interior, to think of their own bodies in parts, and the importance and possibility of the self-improvement of the body through its manipulation — these were all important to the profile and profession of the surgeon. There were, of course, more complex motives. Anatomical displays had been touring across Europe, and across its class divisions, for a long time; but in this context the role of the surgeon was a part of a larger enterprise by an industrialized and bureaucratized society to structure a mass population. As Burmeister puts it, by the time the anatomical museum reached its largest audiences, it was a further reinforcement of a society that had become "thoroughly classed, raced, and unambiguously sexed."[13] Foucault's medical gaze had created, through the presentation of the body in pieces, the body as system. Although greater concern for personal health, hygiene, and diet, and a more rational understanding of medicine showed social concern and sympathy, it was also a means to make the "new" medical profession indispensable.*

This relationship persists in Body Worlds. Von Hagens advocates democratizing anatomy, and in effect demystifying what the anatomist does.[14] He believes that the informed citizen will enter into a better partnership with the doctor and will also in general take better care of the body — just as in the Victorian museum — by imagining the body in its parts. In this respect the pedagogical imperative has a clear historical through-line. This is also true with respect to von Hagens' statements about the aesthetic manipulation of the specimen as a means to educate.[15] The Victorian museum's emphasis on the artifice and beauty of the model can be seen as a distancing device for an audience to whom the manipulation of the body from the inside out, to healing effect, was new. *The Anatomical Venus* was life-like, and artificial, and (at least so advertised) beautiful. The second and third of these attributes — artificiality and beauty — made the likeness to life acceptable.

Body Worlds to some extent represents a similar process of reception, the plastinated body replacing the wax model. If so, however, this does not explain

**On the body, science, race, sexuality, see Bennett,* The Birth of the Museum*; Kemp and Wallace,* Spectacular Bodies*; and Jordanova,* Sexual Visions*; Goodall,* Performance and Evolution in the Age of Darwin*; Stepan,* The Idea of Race in Science*; Lindfors, ed.,* Africans on Stage*; and Gilman,* Difference and Pathology.

***Opposite:* Figure 4: Bill advertising Kahn's anatomical museum in London, circa 1857, illustrating the proximity of models and preparations connecting human development, moral pedagogy, and "malformations and monstrosities." John Johnson Collection, Waxworks 3 (31) and (38), Bodleian Library, University of Oxford.**

the troublesome exceptions described above: intrusions of the donor's individual distinguishing features; non-pedagogical, grotesque manipulations of the body; and the exhibition of deformity. There were, however, other attributes of the Victorian anatomical museum; these are also instructive when placed beside the contemporary example:

1. The display of "authentic" specimens. The Victorian anatomical museum's re-creations of the human body were surrounded, in most cases, by more authentic specimens—authentic within the bounds of decay and the techniques of preservation at the time. Skeletal remains, as well as tissue samples preserved in spirits, were common.
2. The display of "named" specimens. At least some of these remains were identified and personalized by fame or notoriety, either for an action (a murderer), or a state of being (a giant, a dwarf). By way of example, the 1842 inventory for George Langstaff's collection, a wealthy private collector whose access to "the Workhouse of Cripplegate Parish, the out-poor of a very large district, and my own private practice, have opened to me additional sources," included the "skeleton of a Negro" specifically catalogued, as well as "a sequestrum ... from the jaw of the late Mr. Joseph Grimaldi, the celebrated clown." The advertisement for Rackstrow's popular touring exhibition of 1782 included "An exceeding fine Skeleton of a Woman—They were the bones of Mary Edmonton, who was executed in the year 1759, for the murder of her Aunt, Susanna Walker"; "A figure of the late Mr. Bamford, the Staffordshire Giant, 7 feet 2 inches high, moulded from him when alive"; and "A Figure of the Late Mr. Coan, the Norfolk Dwarf, moulded from him when alive—he was 6 and 30 years of age when he died, measuring 3 feet 6 inches height, and was perfectly straight and well-proportioned." These identified figures, one "real Preparation" and two "resembling Life," were displayed along with an Egyptian mummy, wax figures of the Royal Family, a rhinoceros hide, and snowshoes from Lapland.[16]
3. The display of racial difference. After 1850, many of the touring anatomical museums included a separate hallway exhibiting the "Museum of All Nations," which displayed models (and some skeletal remains) of the hierarchy of races then being popularized in scientific discourse. Typically the descriptions, the imagery, and the remains/models were grouped around the ascending order of the separate races of humanity:

> Dr Kahn has also recently added a collection of wax models, taken from life, and exhibiting all the principal varieties of the human race. This collection, which is entirely unrivaled for accuracy, elegance and scientific research, is, moreover, illustrated by popular lectures explanatory of the influences exerted on the human frame by difference of climate, food, and habits of life, together with the modifications caused by peculiarity of temperament, age, and mode of education. Here will be found in one view the graceful Circassian, the angular Copt, the erect European, and the ungainly and ill-poised Ethiopian.[17]

4. The display of physical deformity. Displays included physical aberrations, the common currency of a number of popular forms of exhibition. But in addition to the giant or the dwarf, anatomical displays had a license to display more extreme examples in the name of pedagogy. Popular and widespread examples included documented cases of a horn-like tumor growing from the head (models of "Mr Duval" and "Madame Dimanche") found in Reimers' museum as well as Kahn's, where the following is also listed: "A figure representing the Chinese, A Kin, in whose chest the head and neck of a foetus were found after death, the body being suspended, as in the model."[18]
5. The display of sexuality in the context of "perversion." There were few places where sexuality could be discussed among groups of young Victorians. Whatever the original intentions of their proliferation, anatomical museums quickly filled this need in a controlled manner. Displays and separate lectures and pamphlets appeared on venereal disease, on the reproductive organs, on sexual practice, and on pregnancy and childbirth, that were at the time both praised as an effort to improve the lives of the working class, and decried as a means to corrupt it. In general it taught by negative example, emphasizing the deformities of disease and immoral practices (see figures 3 and 4, illustrations of advertisements).[19] The "Effects of Tight Lacing," for example, concerned the results of ignoring and hiding pregnancy.

Many aspects of a tradition do not survive the passage of time, and I would not wish to suggest that Body Worlds has overtly preserved each and every one of these five alternative features of its Victorian antecedent. And yet, in some respects the experience of attending the contemporary exhibition does carry the imprint of that earlier experience, both in its presentation and in our reception (we are, after all, the descendents of those first visitors). When I review my visit to Body Worlds with the above information in mind, I can see those remnants, in ways that may explain some of the more contentious displays. The effect of confronting a "named" specimen in the Victorian museum remains, in those previously mentioned recognitions of past existence—the tattoo, the crooked tooth. The depiction of race remains, not quite identifiable in the plastinates, fore-grounded by the articles that preceded the exhibition detailing negative accusations against Body Worlds concerning the alleged sources for its bodies (in the former Soviet Union).[20] Something like the effect of witnessing Victorian displays of the giant and the dwarf are quite clearly approximated in Body Worlds as well: a horizontally sliced sagittal specimen was laid out in a glass case separated to create the impression of the remains of a man at least eighteen feet tall; a whole body plastinate was intentionally expanded to gigantic proportions and placed on an oversized bicycle, that intention stated in the exhibit (without pedagogical defense); a plastinate of a child's circulatory system, the rest of his body removed, created a miniature of striking detail.

These possible vestiges of a past tradition may be the result either of my own imposition of a knowledge of the past on the present or, more interestingly, my own cultural inheritance. However, the remaining features of the Victorian museum noted above — displays of "authentic" specimens and of sexuality — more obviously remain intact in Body Worlds, in one case pervasive, in the other segregated, apparently an *homage.*

The Victorian anatomical museum focused attention first of all on the *Anatomical Venus* as a life-like display of artificiality; but these models were surrounded by specimens advertised as "real." This was a kind of authenticity-by-association; and while its persistence in Body Worlds is unarguable, the effect has changed considerably, because the raw material for the artifice *is* the authentic specimen. There is no longer a degree of distance between the two; the plastinate display is at once an original and a manipulation, both authentic and artificial. The Victorian display of authentic specimens looked nothing like the individual donor; they were skeletal remains, or partial relics preserved in spirits. A wax model, however life-like, was not a "real" body, and was looked at quite differently. But in Body Worlds, the spectator sees the artist, the teacher, and the original body all at once. I would argue that this explains in some measure my own response to the more contentious displays. Where the work of the anatomist more closely approximates a perceived state of being in the body, however fictitious — a game of chess or a basketball player — there is not as much to set the artist and the raw material at odds with each other. But when the artist interferes more radically, and yet leaves behind vestiges of the individual, the results seem to me less like a body in arrested motion to some pedagogical purpose, and more like a body in violent contest with the artist for control. I would suggest that some of the elements of race, individuality, and deformity evident in the displays of Body Worlds are obscured by — but occasionally escape from — a technology that employs as its raw material not wax or *papier maché*, but the body itself.

On the other hand, there is one display that clearly exposes a direct influence of the older form on the new; it obscures none of these issues, and stands out for that very reason. Toward the end of the exhibition, just past the plastinate in prayer, along a dead-end alcove, was a display devoted to fetal development and childbirth. A central case contained vials showing the development of the fetus from four days through the sixth week. Along one wall, plastinates showed the fetus in healthy, normal development from the seventeenth through the thirty-third weeks. Along the other wall were contrasting specimens showing fetal birth defects, including Hydrocephaly, Ancephaly combined with a cleft palate, *spina bifida*, as well as a fetus in which the internal organs had grown on the outside, and conjoined twins.[21] These figures were posed with legs and arms crossed, with the exception of the figure of *spina bifida*, which was lying face down. The twins showed signs of bodily hair. At the end of these three parallel displays was a whole body plastinate of a *Reclining*

Woman in the Eighth Month of Pregnancy. She was posed leaning on her right arm, her left reaching up and behind her head. Her lips were slightly separated, and appeared red. Her belly had been cut open to reveal the fetus inside.

I describe this exhibition in detail because it was in structure and substance very much like exhibitions in the Victorian museums, including the authentic specimens depicting the growth of the fetus, the display of the deformities of pregnancy and birth, and even (especially) the positioning of the female model, reclining with one arm raised behind the head.[22] Indeed, the two have so much in common that it would be difficult to believe there is no direct connection.[23] (See figures 3 and 4, reproduced bills of advertisement). And if so, this may explain why, of all the parts of the Body Worlds exhibition, this seemed to stand out itself as an aberration. The positioning of the woman is an aesthetic reference to the past, and not in any stated way pedagogical. The fetuses are in fact the only whole body exhibits showing pathological deformity in the original body. Here in this context, they focus attention on maternity and on birth as somehow dangerous. In this hallway the questions concerning Body Worlds raised above, but obscured in the rest of the exhibition, seem clearly exposed: the presence of authenticity in the very real, very specific deformities; the voyeurism inherent in the exhibition of physical deformity itself; and the proximity of deformity and sexuality as if the two were related.

The extent of the influence of the past on the present should always be treated with suspicion, however similar they seem in the comparison. Similar circumstances may give rise to similar structures without any direct influence. Eyewitnesses (such as me) may impose the structures of the past on the present in their description and interpretation. Similar patterns of activity may have been created by entirely different circumstances. But it is also true that old habits die hard, and that even the more arcane and extreme habits, however discredited, remain embedded in the culture, nearly invisible. I believe that all of these prospects pertain to the culture of Body Worlds, and in particular to the complex relationship between its pedagogical and its aesthetic intentions. More broadly, if the Victorian anatomical museum devoted itself to the creation of an "anatomical world view," and to the advocacy of both self-control and the authorization of the professional to manipulate the body, then these messages are just as clearly and forcefully stated now.

Erasure of Tradition: Body Worlds 2 in 2006

The persistence of this dominant message does not, however, finally explain the extreme manipulations and aberrations I witnessed at Body Worlds in London. Surely they undermine the pedagogy, and betray an aesthetic grotesque at odds with the norm promoted in the rest of the exhibition. It is important to remember, however, that the year was 2002, the place was east

London, and von Hagens did not yet have a relationship with the large public museums that he has today. It makes some sense that the persistent elements of his nineteenth-century predecessors that I found so troublesome also caused a broader institutional concern. A visit to a more recent incarnation of Body Worlds— Body Worlds 2 (in my case at the Ontario Science Centre, Toronto, Canada, in February 2006)—confirms this.[24] The unsettling and non-pedagogical features of the Body Worlds I visited in London were gone. The wall of newspaper clippings had been replaced by an introduction to anatomy. A prominent wall display outlined how and why to donate one's body for plastination as a means for public education. Although there were certainly some unusual whole body plastinates—*Exploded Man* resembled a puppet, and *The Drawer Man* a piece of furniture — the pedagogical imperative was ubiquitous, the aesthetic almost nowhere to be seen. There was no sign of *The Mystical Plastinate*, nor anything like it. And, although there was still a separate alcove for the exhibition on childbirth, only normal fetal development was on display; and even for these specimens, a sign forcefully stated that they were part of a collection of antiques, and not plastinated. There was a whole body plastinate of a pregnant woman, but she was standing in a non-descript position in a glass case; through a relatively discrete incision (by comparison to the London model) the visitor could (barely) glimpse a five-month-old fetus. The accompanying literature emphasized that the woman had died of cancer, and that in such instances a fetus of that age could not survive. The care taken to allay any fears of abuse, inappropriate manipulation of the dead, or unwillingness of any individual to be present in this position after death, contrasted entirely with its London predecessor. Von Hagens, if the comparison holds, excised some of the more contentious museum traditions discussed above — some of which he had earlier taken care to imitate.

The anatomical museum has always been a site of contest over the body between oneself and another — between the individual and the anatomist/professional. It was all well and good, in the nineteenth century, to advocate that working men and women "know themselves" and take control over their own bodies; but the deeper message was to respect and trust the medical profession. The nineteenth-century medical establishment's support for anatomical museums turned to dismissal and outrage whenever these entrepreneurial curators stepped away from this one clear message — when, for example, they began to offer real advice or treatment, most often on sexual matters. Then they were attacked as obscene, the curators declared "quacks."[25] Similarly, Body Worlds as it existed in London in 2002 asked difficult questions of its visitors— about power, respect, control, and trust — that made it unpalatable in some quarters of institutional culture. Since then Body Worlds exhibitions have become significant events at large public museums, important sources of income and an expanded clientele. An accommodation has been reached between the medical/scientific establishment and the popular anatomical museum, in a *reprise* of the mid–Victorian age. As a result, less of the past is visible now. Vestiges of earlier traditions remain, of course; they cannot entirely be erased.

Notes

1. Reimers, *J. W. Reimers's Gallery of All Nations and Anatomical Museum.*
2. Institute for Plastination.
3. Finn, "Art of Controversy"; Bonfante, "The Anatomy of Death"; Moore, "Why I must bring death"; "'Dr Frankenstein' defends"; Leitch, "Shaken but not stirred"; Bauer, "Plastinated Specimens," 222–23.
4. Gunther von Hagens, "Anatomy and Plastination," 9–15, 20–21.
5. Institute for Plastination.
6. Ibid.
7. Ibid.
8. Calvert and Cohen, "Autopsy showman has built up stockpile of 500 bodies"; McKenzie, "Opinion/Interview"; Posner, "'We must embrace death to understand life.'"
9. Levi, "On Microhistory," 93–113, 109.
10. Reimers, *J. W. Reimers's Gallery*, 26; Dreyfus, "The Anatomical Models of Dr. Auzoux," 60–62.
11. "Dr. Kahn's Anatomical Exhibition."
12. Burmeister, "Popular Anatomical Museums," passim; Sappol, *A Traffic in Dead Bodies*, passim; Richardson, *Death, Dissection and the Destitute*, passim; Foucault, *The Birth of the Clinic*, passim.
13. Burmeister, 132.
14. Institute for Plastination.
15. Ibid.
16. Langstaff, *Catalogue*; Rackstrow, *A descriptive catalogue.*
17. Kahn (London, n.d.).
18. Reimers (London), items 164–165, 15; Kahn (London, n.d.), 24–25; Bogdan, *Freak show*, passim; Thomson, ed. *Freakery*, passim.
19. Burmeister, 132–9.
20. von Hagens, "Anatomy," 9.
21. Whalley, *Gunther von Hagens'*, 127 (for photographs); Hudson, "Commentary—Moved by mother and child" (for a positive response).
22. Kahn, Catalogue of Dr. Kahn's (London, 1851), 33 (for illustration).
23. Burmeister, 132–39; Kahn's Catalogue (London, n.d.)
24. Johnson, "My Body in Plastic," 74–80; Ghenu, "Fake plastic bodies."
25. Burmeister, Ch.5–6, 179*ff*.

Bibliography

Bauer, Axel W. "Plastinated Specimens and Their Presentation in Museums—A Theoretical and Bioethical Retrospective on a Media Event." In Whalley, *Gunther von Hagens'*, 222–234.

Bennett, Tony. *The Birth of the Museum: History, Theory, Politics.* New York: Routledge, 1995.

Bogdan, Robert. *Freak Show: Presenting Human Oddities for Amusement and Profit.* Chicago: University of Chicago Press, 1988.

Bonfante, Jordan. "The Anatomy of death: An exhibition of preserved cadavers in Germany is provoking fascination, skepticism and outrage." *Time*, December 15, 1997.

Burke, Peter, ed. *New Perspectives on Historical Writing.* Cambridge: Polity, 1991.

Burmeister, Maritha Rene. "Popular Anatomical Museums in Nineteenth-Century England." PhD diss., State University of New Jersey at Rutgers, 2000.

Calvert, Jonathan, and David Cohen. "Autopsy showman has built up stockpile of 500 bodies." *The Sunday Times*, November 24, 2002.

Drefus, Mark. "The Anatomical models of Dr. Auzoux," *Medical Heritage* 2, no. 1 (1986): 60–62.

"Dr. Kahn's Anatomical Exhibition." *London Era*, March 30, 1851, 11.

Finn, Peter. "Art of controversy: anatomy as entertainment." *Herald Tribune International*, March 6, 2001.

Foucault, Michel. *The Birth of the Clinic: An Archaeology of Medical Perception*. New York: Vintage, 1975.

Ghenu, Mike. "Fake plastic bodies: A controversial exhibit excites, but fails to ruffle visitors' feathers." *Varsity* (University of Toronto), October 3, 2005.

Gilman, Sander L. *Difference and Pathology: Stereotypes of Sexuality, Race and Madness*. Ithaca: Cornell University Press, 1985.

Goodall, Jane R. *Performance and Evolution in the Age of Darwin: Out of the Natural Order*. London: Routledge, 2002.

Hudson, Christopher. "Commentary—moved by mother and child." *Evening Standard*, March 22, 2002.

Institute for Plastination. *Gunther von Hagens' Body Worlds: The Anatomical Exhibition of Real Human Bodies*. DVD. Heidelberg: 2005. Voiceover translation cited in essay.

Johnson, Stephen. "My Body in Plastic: A Tourist Goes to Body Worlds," *Canadian Theatre Review* 127 (2006): 74–80.

Jordanova, Ludmilla. *Sexual Visions: Images of Gender in Science and Medicine Between the Eighteenth and Twentieth Centuries*. New York: Harvester Wheatsheaf, 1989.

Kahn, Joseph. *Catalogue of Dr. Kahn's Anatomical Museum*. London, 1851.

_____. *Catalogue of Doctor Kahn's Anatomical Museum, now exhibiting at the Portland Gallery, 316 Regent Street, Opposite the Polytechnic Institution*. London, n.d.

_____. *Catalogue of Dr. Kahn's Celebrated Anatomical Museum ... now exhibiting in the Music Hall, Nelson Street, Newcastle-on-Tyne*. Newcastle, n.d.

Kemp, Martin, and Marina Wallace. *Spectacular Bodies: The Art and Science of the Human Body from Leonardo to Now*. Berkeley: University of California Press, 2000.

Langstaff, George. *Catalogue of the Preparations Illustrative of Normal, Abnormal, and Morbid Structure, Human and Comparative, Constituting the Anatomical Museum of George Langstaff*. London: John Churchill, 1842.

Leitch, Luke. "Shaken but not stirred: 1.5m expected to see skinned corpses exhibit." *Evening Standard*, March 22, 2002.

Levi, Giovanni. "On Microhistory." In *New Perspectives on Historical Writing*, edited by Peter Burke, 97–199. Cambridge: Polity, 1991.

Lindfors, Bernth, editor. *Africans on Stage: Studies in Ethnological Show Business*. Bloomington: Indiana University Press, 1999.

Moore, Wendy. "Why I must bring death to the people: Controversial anatomist defends his display of corpses as Alder Hey families protest and a legal challenge is prepared." *The Observer*, March 10, 2002.

Posner, Michael. "'We must embrace death to understand life': Amid charges that he violates the dignity of the human body, Gunther von Hagens has taken his show of polymer-filled organs and skin-stripped cadavers to countries around the world..." *The Globe and Mail*, June 25, 2005.

Rackstrow. *A descriptive catalogue (giving a full explanation) of Rackstrow's Museum: consisting of a large, and very valuable collection of most curious Anatomical Figures, and real Preparations; also Figures resembling Life; with a great variety of Natural and Artificial Curiosities; to be seen at No. 197 Fleet-Street, between Chancery-lane and Temple Bar*. London, 1782.

Reimers, J. W. *J. W. Reimers's Gallery of All Nations and Anatomical Museum*. London, 1853.

_____. *J. W. Reimers's Gallery of All Nations and Anatomical Museum*. Leeds, 1853.

Richardson, Ruth. *Death, Dissection, and the Destitute*. New York: Routledge & Kegan Paul, 1987.

Sappol, Michael. *A Traffic of Dead Bodies: Anatomy and Embodied Social Identity in Nineteenth-Century America*. Chichester: Princeton University Press, 2002.

Stepan, Nancy. *The Idea of Race in Science: Great Britain, 1800–1960*. London: Archon Books, 1982.

Thomson, Rosemarie Garland, ed. *Freakery: Cultural Spectacles of the Extraordinary Body*. New York: New York University Press, 1996.

von Hagens, Gunther. "Anatomy and Plastination." In Whalley, *Gunther von Hagens'*, 9–37.

Whalley, Angelina, ed. *Gunther von Hagens' Body Worlds: The Anatomical Exhibition of Real Human Bodies.* Translated by Francis Kelly. Heidelberg: Institut für Plastination, 2005.

Illuminating the Soul: Panopticism and the Freak Show

Pedro Ponce

Without the benefit of instructional signage, audio commentary by its creator, and literary cameos by the likes of Leonardo da Vinci and Friedrich Nietzsche, visitors to Gunther von Hagens' Body Worlds could easily mistake it for the kind of grotesque spectacle typical of the carnival midway. The exhibition has been compared "to a Victorian-era freak show," despite its stated goal of educating the public on the nature of anatomy.[1] The comparison resonates even more if one considers the ambivalence informing popular culture during the nineteenth and early twentieth centuries. According to Rachel Adams, the spectacle of the sideshow freak both flouted and maintained the developing boundary between high and low culture: "Defining themselves against the disorderly clutter of popular entertainment, institutions such as museums, parks, and zoos sought to enlighten their visitors through the strict organization of space and regulation of behavior."[2]

While partaking of the latter efforts at distinction, one cannot deny that von Hagens' exhibits also share in the "disorderly" amusements of the sideshow. *The Winged Man*—whose facial muscles have been spread open to resemble wings—suggests the conjoined bodies of Siamese twins; the white hat atop one of the exhibit's heads invites viewers to recognize the humanity within this gruesome display, eliciting the kind of attraction and repulsion experienced by sideshow patrons.* Several plastinates are exhibited in everyday postures, such as playing chess or basketball, while their tissue and organs are exposed with unsettling transparency. If viewers identify with the humanity of these specimens, they must also identify with the substance of this humanity—a fragile hodgepodge of muscle, bone, and nerve tissues. At the same time that von Hagens elicits wonder at the intricate mechanisms of the body, he is also questioning

**All observations on Body Worlds in this article are based on attendance at the exhibit in Charlotte, North Carolina, in September 2007. The dual reaction elicited from freak show audiences is described by Grosz, 57.*

their integrity. This contradiction is apparent in the signs welcoming visitors to the exhibit, assuring them that what they are about to experience will help them to "better understand the human body and its functions" even as it reveals the "vulnerability of the human body."

What is at stake in this vulnerability is the integrity of the individual self modeled by embodiment. According to Susan Stewart, the mouth, anus, and other openings on the human body constitute one's sense of "what is the body and what is not"; when these thresholds are exceeded, the grotesque is embodied in the "scattering and redistribution of bodily parts."[3] The inversions cited by Stewart — a mummer with enormous fake breasts, the class clown who substitutes his exposed rear end for a face in a car window[4] — seem a far cry from von Hagens' sober clinical rhetoric. Nevertheless, reading Body Worlds as part of sideshow grotesque tradition is important for understanding the spectacle that underpins the exhibit's ostensible goals, as well as the extent to which all such spectacles regulate the way audiences see themselves.

At the conclusion of Body Worlds, a smiling von Hagens is pictured next to some final words to the viewer, in which the doctor explains how plastination "transforms the body, an object of individual mourning, into an object of reverence, learning, enlightenment, and appreciation." But the idealism of this "blowoff" (sideshow parlance for the exit of an exhibit[5]) is complicated by the spectacle it purports to explain. If von Hagens desires to remove the taboo from confronting death, he does so by raising the specter of another, as described by Susan Stewart: "What is both inside and outside the body (feces, spittle, urine, menstrual blood, etc.) tends to become taboo because of its ambiguous and anomalous status."[6] Despite its packaging in educative discourse, Body Worlds confronts viewers with the anomaly of insides displayed as outsides, individual bodies composed of anonymous, insentient fragments. It is just this ambiguity that the so-called freakish body presents within the less exalted world of the sideshow. In his description of viewing Siamese twins, Leslie Fielder arguably articulates precisely what is so disturbing about confronting von Hagens' purportedly edifying anatomical models:

> standing before Siamese Twins, the beholder sees them looking not only at each other, but — both at once — at him. And for an instant, it may seem to him that he is a third brother, bound to the pair before him by an invisible bond; so that the distinction between audience and exhibit, we and them, normal and Freak, is revealed as an illusion, desperately, perhaps even necessarily, defended, but untenable in the end.[7]

It is this sense of connection — between spectacle and viewer — that is provoked by Body Worlds as well as its precedents in early American mass culture. While criticism of the sideshow as cultural phenomenon frequently explores the sort of disorientation described by Fielder, von Hagens' appropriation of the freakish highlights its ambivalent relationship to the viewing audience — reinforcing boundaries demarcating the "normal" even as it unsettles these boundaries.

"'Freak' is a way of thinking," observes Robert Bogdan, "of presenting, a set of practices, an institution — not a characteristic of an individual."[8] Ultimately, the display of the sideshow exhibit says more about how otherness is interpreted than about the nature of the purported anomalies presented.

Recent criticism has emphasized the salutary effects of exposing turn-of-the-century consumers to puzzles such as P.T. Barnum's Feegee Mermaid and other faked curiosities, exercising the public's capacity for reason at a time when significant cultural questions threatened to overwhelm it. While acknowledging the "dehumanizing" qualities of Barnum's What Is It? exhibit — a humanoid hybrid portrayed by an African-American man — James W. Cook discerns the complex social forces at work in Barnum's spectacle and in the audiences it fascinated: "Barnum offered only a range of possibilities. But in doing so, he also initiated something else: a series of public conversations in which the potentially divisive issues of sectional politics, racial science, and social respectability began to serve as topics of popular amusement."[9] Sideshow entertainer Doug Higley echoes this observation, as quoted in Joe Nickell's more contemporary study:

> When Warner Bros. does a movie featuring a Lost Jungle City, do they put on the poster, "Not real! Untrue! This is Fiction!"? ... I look at every show I do or every gaff I create for others as my own little movie. The fiction is the fun, the mystery is the value, the fantasy is the goal. Each is designed to work the brain — crinkle the brow and crack a smile, and, yes, in some cases even educate and inform.[10]

Nevertheless, read as a sideshow, Body Worlds suggests the ways in which the freakish spectacle "distances the viewer, and thereby ... 'normalizes' the viewer as much as it marks the freak as an aberration."[11] Von Hagens employs equal parts pedagogy and pandering to engage viewers along a boundary easily overlooked amid the more sensational breaches— man versus woman, self versus other, human versus animal — played out on the midway: diversion versus discipline.

The contemporary sideshow is informed by a long and eclectic line of popular entertainments going as far back as ancient Egypt.[12] According to Andrea Stulman Dennett, the first recorded business of exhibiting freaks occurred in thirteenth-century England, and such displays are even referenced by William Shakespeare in *The Tempest*.[13] Among the entertainments that contributed to the sideshow were menageries, collections of animals for display such as the Chinese emperor Wen's "'garden of intelligence'" from the twelfth century B.C., "wherein animals from the different provinces of the empire were exhibited."[14] In their many permutations, such displays represent an attempt at naming, classifying, and consequently controlling the other. Stewart connects the sideshow to this colonialist impulse when she observes, "On display, the freak represents the naming of the frontier and the assurance that the wilderness, the outside, is now territory."[15] For all its appeal to audiences' desire for a taste of the seamy and exotic, the typical sideshow was a rigorously controlled entertainment;

many giants, for example, were legally forbidden to be measured, the better to exaggerate their abnormal height, only one of numerous ways in which exhibits were manipulated to draw curious crowds.[16] The crowds themselves were just as manipulated: by shills posing as audience members lining up eagerly for tickets; by banners exaggerating the authenticity of the exhibits inside; and by promises of even greater attractions if one paid extra.[17] As Stewart suggests, China and Coney Island may not be that far apart after all. In both cases, the other is rendered visible within a particular cultural framework.

The power exerted through visibility is explored by Michel Foucault in *Discipline and Punish.* The panopticon, template for the modern prison, operates by isolating individual inmates under the gaze of a central observer, the latter unseen. The purpose of this design is "to induce in the inmate a state of conscious and permanent visibility that assures the automatic functioning of power."[18] In a particularly apt analogy, Foucault suggests a tantalizing connection between the prison and the sideshow: "one finds in the programme of the Panopticon a similar concern with individualizing observation, with characterization and classification, with the analytical arrangement of space. The Panopticon is a royal menagerie; the animal is replaced by man, individual distribution by specific grouping and the king by the machinery of a furtive power."[19]

If "[v]isibility is a trap," then by rendering individual difference visible, the dynamic of the panopticon renders difference potentially more accessible to regulation.[20] Understood via Foucault, the sideshow invites a reconsideration of how power is distributed within its individual exhibits. The scope of this article does not allow for a full treatment of the sideshow in these terms, but a few examples suggest that von Hagens' is not the only exhibit that can be productively scrutinized as a panopticon. Consider how much of the deceptions involved in making sideshows work depend on reading the body correctly. Nickell cites psychologist and former palmist Ray Hyman describing how a sideshow fortuneteller reads a patron's body for clues to assembling a plausible psychic reading: "If he is on the wrong track the client's reactions—eye movements, papillary dilation, other bodily mannerisms—will warn him. When he is on the right track other reactions will tell him so."[21] An apparent anomaly like a human pin cushion, who can pierce his skin without pain, is not really a biological aberration; rather, he is intimately familiar with where the body is naturally less sensitive to pain.[22]

By the same token, physical anomalies can be reintegrated along familiar cultural lines, however unsettled by the freaks on display. Nickell cites a description of the hirsute Grace Gilbert: "'She was heavily built, and according to legend she was able to perform a man's work on the farm... Yet, like other bearded ladies, she was famed for her ladylike refinement and her skill with needlework and spent much of her spare time making lace.'"[23] And Grace McDaniels, known as the "Mule-Faced Woman," made up for her ugliness by the warmth of her personality—imbued with a maternal, nurturing quality that located her squarely within the gender conventions denied by her appearance: "'She is one of the best-loved women in the carnival business, a homespun, motherly soul, generous to a fault. After five minutes' conversation with Grace you forget all about the strange contours of her face and are aware only of the warm, courageous heart of the woman herself.'"[24] The body, then, is put to use in the sideshow. The sideshow's power dynamic renders "the body as a functional tool" at the same time that it imbues the body with the potential for disintegration and mutability.[25] The question raised by the sideshow is ultimately not who is freak and who is normal, but who is the real inmate?

At the exit from Body Worlds, von Hagens is quoted observing how "[t]he plastinated post-mortal body illuminates the soul ... I hope for BODY WORLDS to be a place of enlightenment and contemplation." But this rhetoric of transcendence is closely tied to the integrity of the body if one consults some of the authorities cited throughout the exhibit. The Greek philosopher Epicurus dispenses a kind of thinking person's tough love when he admonishes the crowd to "[g]et used to the idea that Death should not matter to us, for good and evil are based on sensation. Death, however, is the cessation of all sensation." Nietzsche chimes in next to an exhibit on the *Inner Face*, in which a rotating specimen displays the inside of eye sockets and a mouth obscured by the individual features of the living: "the enlightened, the knowledgeable would say: I am body through and through, nothing more; and the soul is just a word for something on the body." The commentary that accompanies the exhibit reinforces the German philosopher's connection between the body and individual identity: "The external individuality characterizing a person during his or her lifetime is no longer recognizable in the anatomical specimen." In this display, the body becomes a territory that designates the extent—and the limits—of what it is to be human. If, as Foucault has observed, the "soul is the effect and instrument of a political anatomy; the soul is the prison of the body," then von Hagens' rhetorical deployments seek to liberate us from this prison.[26] But the panopticism of the sideshow requires a reframing of the perspective being unsettled; the bodies represented through von Hagens' dissections are visible only within "the frame of the pitchman or barker."[27] The body is rendered as other only to be reintegrated within a discourse that encourages audiences to make the most of their "territory." As with other spectacles that evoke otherness, Body Worlds is less about celebrating individual experiences of the body than about illuminating what passes for normal.

One strategy employed in Body Worlds to this purpose is the appropriation of freakishness to distinguish between ideal specimens of the working body and diseased specimens of failed bodies. Some of the working systems are displayed with faces intact, such as *The Basketball Player* (2002), which presumes to illustrate the normal musculature of an athlete, or *The Chess Player* (1997), whose posture is used to show the structure of the nervous system. To emphasize the "work" accomplished by healthy systems, the exhibit uses mechanistic analogies to explain what different parts of the body ideally do. The liver, for instance, "has a key function in metabolizing carbohydrates, fats and protein and serves as a 'sewage-treatment plant' for the blood, in that it absorbs toxic substances from the blood and transforms them into harmless waste products." Audio commentary on the *Nerve Body* (1999) invites the viewer to think of the nervous system "as a computerized communication system [in which] the brain is like a control center, the spinal cord like a router, and the nerve fibers function as the connecting cable." In striking contrast to some of the healthy specimens on display, *Nerve Body*—which shows the effects of severe spinal arthritis—appears without a face, one eye socket empty. *The Smoker* (1996) is presented with half a face, in keeping with its binary display, one side of the plastinate showing muscle structure, the other showing the discoloration of lung tissue caused by smoking. This posture brings to mind the so-called "half-and-half" display of sideshow hermaphrodites, who split their appearances down the middles of their bodies between exaggerated male and female features.[28]

This distinction, seen in some of the Body Worlds exhibits, is only part of how von Hagens embodies boundaries of the normal while ostensibly celebrating the wonders of individual anatomy. For how can an exhibit like *The Runner* (1997) not be construed as grotesque? This plastinate is designed to show "how tenuously the sinewy muscles are attached to the bones." Its facial features intact, its appearance is nevertheless highly unsettling; viewed in the round, it resembles a skeleton wearing flesh-colored pelts as part of some bizarre, doubtless occult ritual. Other displays of "normal" bodies are similarly manipulated into freakish postures. The sideshow tall man is replaced by Body Worlds' *The Longitudinally Expanded Body* (1996), whose skull is opened up to display its eye socket and tongue, and whose lower parts are similarly cut open and expanded to reveal what is inside the chest cavity, pelvis, and knee joints. *The Organ Man* (2000) is a solicitous instructor to the viewing public, holding his own liver, gallbladder, and stomach.

The riddle of these ostensible anomalies is solved in part if one considers the discourse surrounding the *Frontal 3-D Slice Body*. Hung before the audience in five cross-sections—much like clothes on hangers—the body presents a complex puzzle of body parts in need of assembly. We are assured, however, by the accompanying commentary that this puzzle does indeed have a solution: "The resulting negative and positive forms provide an instructional pattern of complementary protrusions and indentations." In the context of Body Worlds, individual differences matter little. (The *Slice Body* reinforces this impression by a sign that escapes the exhibit's carefully structured anatomical narrative—tattoos are visible on its upper arms.) What matters, according to the bodies on display, is the anatomy we share—fragile, mortal, but ultimately explainable, as long as we remember to make our bodies useful.

Von Hagens would deny that his edifying exhibit is a sideshow, and to a great extent he would be correct. Body Worlds is much closer to the narrative of the museum, which downplays sensationalism in the service of educating the public. However, in the regulated sensationalism of its displays, in the power of its discourse that attempts to work on the bodies of its "inmates" — whether the bodies themselves or the bodies of those who pay to see them — in the way it extols the individual even as individual difference is reduced to an accident of anatomy, Body Worlds can arguably lay claim to the title of Greatest Show on Earth.

Notes

1. Schmidt, "Controversial Cross-Section," 39.
2. Adams, *Sideshow*, 26.
3. Stewart, *On Longing*, 104–105.
4. Ibid., 105.
5. Nickell, *Secrets*, 76, 77.
6. Stewart, 104.
7. Fielder, *Freaks*, 36.
8. Bogdan, *Freak Show*, 10.
9. Cook, *The Arts of Deception*, 138, 139.
10. Nickell, 340–41.
11. Stewart, 109.
12 Nickell, 1.
13. Dennett, *Weird and Wonderful*, 66.
14. Nickell, 299.
15. Stewart, 110.
16. Nickell, 84; Bogdan, 10.
17. Nickell, 66, 78, 146–47.
18. Foucault, *Discipline & Punish*, 201.
19. Ibid., 203.
20. Ibid., 200.
21. Nickell, 269.
22. Ibid., 233–34.
23. Ibid., 154.
24. Ibid., 163.
25 Stewart, 105.
26. Foucault, 30.
27. Stewart, 109.
28. Nickell, 196–97.

Bibliography

Adams, Rachel. *Sideshow U.S.A.* Chicago: University of Chicago Press, 2001.

Bogdan, Robert. *Freak Show: Presenting Human Oddities for Amusement and Profit*. Chicago: University of Chicago Press, 1988.

Cook, James W. *The Arts of Deception: Playing with Fraud in the Age of Barnum*. Cambridge, MA: Harvard University Press, 2001.

Dennett, Andrea Stulman. *Weird and Wonderful: The Dime Museum in America*. New York: New York University Press, 1997.

Fiedler, Leslie. *Freaks: Myths and Images of the Secret Self.* New York: Simon and Schuster, 1978.
Foucault, Michel. *Discipline & Punish: The Birth of the Prison.* Trans. Alan Sheridan. New York: Vintage, 1995.
Grosz, Elizabeth. "Intolerable Ambiguity: Freaks as/at the Limit." In *Freakery: Cultural Spectacles of the Extraordinary Body*, edited by Rosemarie Garland Thomson, 55–66. New York: New York University Press, 1996.
Nickell, Joe. *Secrets of the Sideshow.* Lexington: University Press of Kentucky, 2005.
Schmidt, Natasha. "Controversial Cross-Section." *Index on Censorship* 31, no. 4 (2002): 39.
Stewart, Susan. *On Longing: Narratives of the Miniature, the Gigantic, the Souvenir, the Collection.* Durham: Duke University Press, 2005.

The Amethyst Seal: Anatomy and Identity in Bentham and von Hagens

Patricia Pierson

The 1995 debut of Gunther von Hagens' traveling anatomical exhibition, Body Worlds, drew great controversy that continues even today. Just over a century and a half earlier, though, the celebrated inventor of Utilitarianism, Jeremy Bentham, caused his own scandal when he provided in his will for the dissection, preservation, and display of his own corpse, which currently resides in the South Cloister of University College London and is officially known as the Auto-Icon.[1] As millions of visitors have flocked to the shows, Body Worlds has generated extensive media and now scholarly interest, though until recently few have mentioned the family resemblance between Bentham's and von Hagens' projects. Indeed, even though von Hagens himself coyly hints at a connection to Bentham in his press materials when he observes that he wears a hat in homage to the philosopher, the link between the two men goes beyond the sartorial.[2] It is my argument that both utilize a very specific and storied conception of anatomy to craft indelible and dynamic material narratives of human identity; they are life writing through the body, in the space of the museum, by means of the collection.

While the characteristics of von Hagens' exhibition are widely known, perhaps fewer readers of this volume are familiar with the details of Bentham's Auto-Icon. The Auto-Icon consists of Bentham's skeletonized corpse, which is stuffed and dressed in his own black coat and hat, seated in a favorite chair in a position of contemplation he is said to have assumed when writing. The body is crowned with a wax replica of Bentham's head and is contained in a mahogany-and-glass display case. At one time, the case also contained Bentham's mummified head, which is now kept in cold storage for security and preservation purposes. The head in particular has been difficult to maintain, and has on some occasions been stolen and misplaced. Initially it was housed within the metal ribcage of the assembled skeleton, and then for several years it rested at Bentham's feet. Also contained in the case are an amethyst-colored glass intaglio and Bentham's favorite walking stick, which he called Dapple. The intaglio, a traditional mourning object, features a portrait in profile of Bentham. It is from this object that I take the title of my article, since the intaglio, often used in classical times as a signature stamp, here serves as Bentham's mark or signature upon himself as his own creation — his self-symbol or auto icon. Traditionally, the seal created by such an engraving would render in wax a miniature portrait of its owner, in the same way that the identities of the Auto-Icon and those featured in Body Worlds leave a trace or stamp of themselves, of who they once were, upon their beholders.

Bentham's body arrived in this position as a result of very precise directions in his will. Even before the writing of his last will, he had a long and well-documented history of thinking about the uses of his body after death.[3] In his first will and testament, written as a teenager, he made arrangements for his remains to be donated to science. Toward the end of his life, though, he crafted specific plans for his body that included its complete and public dissection for educational purposes, the mummification of his skull in the "New Zealand" style, and the preservation of his cleaned skeleton for display. Following most of Bentham's wishes, his personal friend and physician Southwood Smith performed a public oration over Bentham's remains at the Web Street Medical College, and a later dissection (perhaps carried out only before medical students, as James Crimmins suggests) was performed by either Smith or another member of the faculty.[4] Bentham's skeleton was then dressed in his choice of clothes and his mummified head was fitted with glass eyes he had selected while still alive. (Crimmins and others note that there is an unsubstantiated legend that Bentham carried these eyes in his clothes pocket in the years before his death, producing them on occasion for his visitors to behold.[5]) For many years the Auto-Icon sat in Smith's home and then in his medical offices, and was later donated to the university where it remains today.

Along with the instructions in his will, Bentham also is thought to have composed an essay about his theory of the Auto-Icon, though its authorship remains in some dispute among contemporary Bentham scholars. In the essay, entitled "Auto-Icon; or Farther Uses of the Dead to the Living," Bentham attempts to insert himself into the contemporary debate over the scarcity of anatomical specimens.* During Bentham's time there was a strong stigma associated with dissection, due to the prevalence of grave robbers ("resurrection men"), a prominent case of a pair of murderers who sold their victims to a physician for dissection, and debates over what eventually became the reforms passed in the Anatomy Act of 1832.[6] In the surprisingly humorous but also

**The debates over dissection current in Bentham's era and the stigma that accrued to surgeons and anatomists are not unlike the contemporary criticisms of von Hagens' project. Concerns about the ethics and legality of the dissection, collecting, and display of human remains is a constant in any popular discussion of these issues, no matter the historical era.*

ingeniously devised fragment, Bentham speculates on the possibility of doing away with burial altogether and allowing the dead to be dissected, preserved, and then displayed in private and public venues for the greater benefit of the individual and society as a whole. His visions for the preservation of the dead and their potential value to the living echo von Hagens' design for the Body Worlds exhibitions and his stated didactic goal of bringing anatomy to a popular audience. Thus the common link between both exhibitions is their treatment of anatomy in a broader sense.

Anatomy and Identity

To consider anatomy is to consider questions of human identity. Anyone who chooses to think on this topic, especially those outside of a traditional scientific context, will study it not so much for practical ends, such as curing disease or alleviating pain, but oftentimes for the purposes of understanding what is to most of us the wholly unreachable mystery of the transformation of death. Yet the notion of anatomy in this sense is not essentially a morbid one; rather, those who ponder it are truly curious about the nature of life and of the living. They are curious about the very central questions of the Humanities: what it means to be human, and what are the nature and scope of human dignity.

Freed from the responsibility of learning and understanding systematically or logically as a scientist would, those who pursue the lay study of anatomy are characterized by their wonder, marvel, and curiosity. For these reasons, Body Worlds has captured the popular imagination around the globe. The affective impact of the exhibition plays upon the inexpressible and unnamable sense of dread Sigmund Freud describes in his analyses of the Uncanny and the automaton.[7] There is something inherent in the moment of viewing another human being — a subject — suddenly as an object, to look without being looked back at, that is both thrilling and profoundly unsettling. The preserved remains inhabit a liminal space between life and death, similar to the automaton that is brought to a semblance of life in Hoffmann's story "The Sandman," which inspired Freud's theory.[8] Similarly, the Auto-Icon replicates the dress and demeanor of Bentham as he was in life, giving the beholder the anachronistic vision of being "eye to eye" with a celebrity of the nineteenth century while standing firmly in the twenty-first. Uncanniness in both cases arises from the sight that really should not be seen, or the sight that is strange or shocking to see.

In viewing the specimens, it becomes clear that something is familiar but also is made strange, as if we are looking at ourselves for the first time in an antique mirror whose partially oxidized silvering renders it simultaneously both a reflective surface and a window. Indeed, part of what has been controversial about von Hagens' technique is not the plastination or dissection of the specimens but instead how they are posed as if pursuing pleasures they might have known in life, such as pirouetting on pointe, horseback riding, or skateboarding. The sense of discomfort or uncanniness evoked by the exhibition is not only due to the living's proximity to the dead but also to the way the dead are, doll-like, posed to imitate the living as doubles from the inside out. The horror of Body Worlds is the creeping realization that viewers and specimens share the same museum space, but are there under wholly different circumstances— some as figures in a vast collection of *natures mortes* and some as consumers of that display.

For children, however, who are comfortable moving among the worlds of animatronic dolls, virtual-reality video games, and make-believe, Body Worlds is pure fascination. It is no accident that the children I saw during my visit to Body Worlds 2 in Los Angeles in 2005 were exclaiming, chattering, and barely able to hold themselves back from touching the specimens, while the adults accompanying them expressed either hushed reverence or defensive gallows humor. This illustrates how both Body Worlds and the Auto-Icon perpetrate the transgressive act of keeping the dead out of their proper place and in the midst of public life. They violate what Freud termed the taboo upon the dead, which insists that the living and their ancestors be kept apart.[9] These exhibitions prey upon what Bentham rightly and sensitively calls in his Auto-Icon essay a "primitive horror" of seeing the dead so intimately.[10]

These are two cases of the concept of anatomy writ large, and demonstrate the significance of anatomy in a theoretical sense to the understanding of human identity. From the Greek for "to cut up" or sometimes, "to cut stone," anatomy can represent the body or regions of it; dissections of the body in order to understand it as a whole; the science (systematized knowledge) of the body; a mummy or a "walking skeleton"; and finally, a treatise written about the body or any topic of a philosophical bent. To anatomize can mean literally to dissect and also, figuratively, to analyze. Body Worlds and the Auto-Icon yoke together these meanings in the figures of their displays, becoming a thing that is dissected by the hand as well as a thing that is analyzed by the mind and represented in the written word.

In a literary sense, the anatomy is a Renaissance love poem, often a sonnet, written as a catalog of the beloved's features but usually only of the face.[11] The term "anatomy" in these ways combines the several themes of human identity and the analysis and writing of identity. I would suggest the intaglio—the amethyst portrait seal—as the metaphor for the particular treatment of identity we see in both Body Worlds and the Auto-Icon. Although the Auto-Icon is clearly a case of the reinscription of an identity already well known to the beholder, while in Body Worlds von Hagens attempts to conceal the identities of the bodies on display, both underscore the human yearning for an understanding of the lives of others. Both spectacles draw in the viewer who wants to know what life was like for that subject when he was alive and living in the body now on display. The intaglio, whose original purpose was to mark a signature or official seal on a document, here serves as a figure for identity and also for printing or writing. It is the possibility of the moment of inscription of one human being's life and image upon the face and psyche of another, the seal of our own psychic transformation upon viewing these preserved remains.

Bodies in the Museum

Both Body Worlds and the Auto-Icon shocked the public in their respective eras due to their treatment of the human body. The special cases that arise when the index of the most personal — the body — is displayed in a most public site — the museum or a public building such as the university — raise questions we would not normally consider about the proper place of the dead, and, more disconcertingly, the proper use of the dead. The consensual display of individual identity, always directed outward at unknown yet anticipated others, calls for a radical reconsideration of the proper subject of the museum, and upon deeper consideration, such instances bring the viewer back to the very original sights of the early *Wunderkammern*, which were displays of the manipulated, posed, preserved, and narrativized human form.

Von Hagens' project, especially, has an ancestor in the museum work of Dutch anatomist Frederik Ruysch (1638–1731) of Leiden, whose specimens were often arranged in tableaux illustrating the *vanitas mundi*.[12] Ruysch's specimens were bought by pioneering collector Peter the Great of Russia, and became the foundation of his *Kunstkamera*. Ruysch, chief anatomist and coroner of Leiden, a pathologist, collector, scholar, and artist, helped to characterize the early museum as a site for the study of the human form. His work was anticipatory of Bentham's Auto-Icon and of Body Worlds, and in this sense what Bentham and von Hagens effect is a resurrection of the historical practices of the museum. While Bentham has provided for himself to become part of a personal collection, housed in his own private cabinet of wonder, Body Worlds is a vast and seemingly exhaustive traveling collection of specimens, brought together by the singular, eccentric personality of von Hagens. Both collections bring their viewers back into a museum space of times past, when the proper subject for the museum was not art but man himself and questions about his place in the larger world.

What is today known as the natural history museum is perhaps the twenty-first century cognate of the cabinet of curiosity, cabinet of wonder, or *Wunderkammer*. The modern museum, a systematized and ordered accretion of specimens and objects, took shape over the last five hundred years from its early incarnations as small private or royal collections. Often displayed in an elaborately inlaid wooden cabinet or *Kunstschrank* ("art cupboard"), or in densely cluttered, dedicated rooms, these *Wunderkammern* organized diverse species of the known and recently discovered world into a single domestic space. The Cabinet of Wonder often demonstrated the possibility of this very quality of regarding the world with an attitude of curiosity, wonder, or awe. The authenticity of the items contained and displayed in the collection was of less significance than their alleged provenance, as the objects were collected on journeys around the world in a booming age of exploration.

Perhaps the most famous of these private, royal collections belonged to Peter the Great. His *Kunstkammer* contained objects representing the voracious and inquisitive czar's many interests, from anomalies of nature such as a two-headed sheep or a four-legged chicken, to preserved anatomical specimens obtained at home and abroad, to works of culture such as pieces of metal turned on his own lathe. Not surprisingly, the museum also housed "live" exhibits, such as a hermaphrodite, a two-fingered and two-toed servant, and, after his death, the body of Bourgeois, Peter's seven-foot-tall manservant. Stephen Jay Gould explains that such collections were intended to stun or shock the beholder, similar perhaps to the spectacular nature of both the Auto-Icon and Body Worlds.[13] In hopes of improving his collection, Peter forged a close bond with Ruysch, who, renowned for his innovative techniques for embalming even the slightest of veins in anatomical specimens, created elaborate dissections that he preserved in alcohol for all to see. Since his skill was a point of pride for his city, it was a great coup for the Petersburg museum when Ruysch sold his collection to Peter in 1717.

The museum tradition of the Petersburg *Kunstkammer* and the anatomical exhibits of Frederik Ruysch are significant because they call attention to the origins of the collection as a site for the display of anatomical wonders. To think of what is now popularly considered such an august tradition as bound up with the display of skeletons posed in vignettes or teeth pulled by the czar is to consider the collection and its authority in a new light. To read Body Worlds in this context is to understand that von Hagens' project is reactionary rather than radical. He is bringing anatomy, and as some critics would argue, the medical "freak" show, back to the museum as an instructive spectacle for the lay public. His work is controversial precisely because he preserves human bodies, yet this gesture of preservation is consistent with the historical mission of the museum.

As Dragan Kujundzic notes, the museum is a site of preservation — both literally, by placing specimens of nature in preservative fluids— and figuratively.[14] Objects must be preserved in the museum because the collector always anticipates her own death, and must "save" the collection for other eyes and other audiences. If the only possible audience for the collection were we who see it now, there would be no reason to preserve it for others. Thus the museum collection constructs and displays the collector's identity, and also anticipates her future selves. It is a space for the free play of identity, threatening, perhaps, in the case of the Body Worlds exhibition precisely because identity is destabilized and restabilized by the objects collected. We do not know the stories, names, or causes of death of those we observe; we know simply that they once lived and have died. In the gap between what we know and what we imagine, identities are postulated and constructed, and the viewers themselves call into question their own identities and histories.

The collection or museum space is a conservative site that retains what might be otherwise lost, but also creates something new out of the old. This preservation also makes it possible for the museum to become a site of parody where real things, wrenched from their contexts and re-contextualized in an artificial space, appear as their own uncanny and often laughable doubles.[15] This parodic element, wholly ignored yet obviously present in the *Kunstkammer,* is perhaps one explanation for why Body Worlds has often been criticized for demeaning the dead. When the dead are taken out of their proper context — the coffin, laboratory, or urn — they may appear undignified or comical. It was this combination of horror and humor that characterized Ruysch's tableaux in his anatomical museum, but in today's arguably more enlightened context it is rightly no longer socially acceptable to use the dead for affective rather than instructive or ostensibly scientific aims.

The Collection as Autobiographical Discourse

The model for the first museum, according to Susan Stewart, John Elsner, and Roger Cardinal, was Noah's Ark.[16] The fantasy of completeness (or what Cardinal and Elsner call the "pathology of completeness"), of making a replica of the universe in one space, fully replicable (two of each kind) but also contained and thus master-able, is the fantasy of all collectors. To take the world into one vessel and then to have the power to steer that vessel, to move it, and, ultimately, to find a new place for it, is the power of Noah and also the ultimate thrill of play for the collector. Things that are collected might never even be looked at again, but they are there as playthings at the disposal of their owner. As Walter Benjamin observes, the relationship between a collector and his things is one of ownership, so that the use of the object becomes less important than its symbolic significance.[17] The Latin words for wonder—*mirari*—and marvelous—*mirabilis*—share a common root with the word mirror, and all reflect the medieval notion of marveling at the sight of something wonderful.[18] One could also argue that wonder contains within it the notion of marveling at oneself.

Stewart notes that the collection and the museum are separate things, but that the museum is the metaphor *par excellence* for the collection.[19] By arranging objects within space, the narrative of production is replaced by the narrative of collection, in that objects in the collection lose the meaning they once had as things that could potentially be used and are placed in a meaningful series within the collection itself.[20] Stewart claims that as the collection is ahistorical, and erases the history of its objects, the objects' history is replaced by that of the collector. Objects become signs in the collection, placed in chains of signifiers, and their ultimate referent is the interiority of the collector's identity.[21] The collector's identity is shaped by the collection, but the collection also introduces an element of instability into this identity because it "overload[s] the self with signification."[22] The collection stands for the world through metonymy because it replaces "part for whole, item for context," and because its classification creates a sense of space and time that makes the collection into its own world.[23] Thus the collection, while entirely about one person's identity, becomes a story of identity recounted and recountable to multiple audiences.

Jean Baudrillard and Walter Benjamin also posit that the collection tells of the collector's identity. In his essay on books, "Unpacking My Library," Benjamin states that it is the collector who truly lives in his objects.[24] Baudrillard, however, claims the collector's identity is constructed by the collection, as she is but one of a long chain of signifiers made up of her objects. An individual's relationship to objects in a collection removes the objects from the realm of utility and exchange, and places them within the feedback loop of the collector's identity. It is in the completion of the collection that the collector's identity is finalized, but that is also the time of his death.[25] In other words, the collection tells the story of a particular subject, but it is also the cause of the subject's demise. In this way, museum spaces, collections, identity, and death intermingle in a mixture of signification common to both Body Worlds and the Auto-Icon.

Yet how may we consider Body Worlds and the Auto-Icon as material life writing or autobiographies? Baudrillard's somewhat critical view of the collection as a solipsistic discourse finds its counterpart in Mieke Bal's concept of telling objects. For Bal, every collection tells a story, working by a logic similar to narrative in that the collection always begins after the fact: the collector begins acquiring objects, and only later sees that he has formed, in retrospect, a collection. Bal connects this to the way that narrative, with its Aristotelian movements of beginning, middle, and end, plays with the ordering of events, plot, and time. Bal sees a kinship between the "paradoxical" situations of both narratives and collected objects, in that both exist independently of anything else, and thus, in the case of narrative, appear to be constative, or, in other words, stating "facts." However, texts like collections have meanings that are produced *intersubjectively* between reader and writer.[26] Additionally, narratives like collections depend upon a narrating agent and her narration thus becomes filtered through her own subjectivity. So the narrator-agent exists as a linguistic function of a text, while at the same time imposing a sort of subjectivity upon the narrative. This tension between what we could reductively call subjectivity and objectivity is what Bal argues is most compelling about narrative and the collection, and it is also the key to the kinship she finds between them.

In light of the history of the *Wunderkammer* and the body of collection theory, it is hard not to see that the collection is always, at its basis, a story about the self, projected onto space and distributed across several objects. It reveals the tension between "objectivity" and subjectivity that is at the basis of all narrative, for as Bal notes, the role of the focalizer makes all narrative to some extent not merely constative but also subjective and thus fictional. For a narrative about the self to be told in things raises the question of, in turn, the unexpected objectivity in a literal sense of all life writing: stories about the subject are invariably collections of events, actions, and places, just like a narrative collection. Thus objects tell stories and become *telling objects*, blending pure objectivity with the always fictional practice of narration.

The cases of von Hagens and Bentham are the two faces of identity in the anatomical collection: anonymity and celebrity. The specimens in Body Worlds, reputedly eagerly donated by willing subjects, have had their identities erased in favor of what von Hagens has called the "inner face" of their bodies. These donors, if it is true that they have all been such, are happy to let their individual identities be subsumed under the general rubric of von Hagens' project. Current controversies as to the origin of many of von Hagens' cadavers— such as the rumors that he utilizes executed mainland Chinese prisoners, or that he purchases bodies by questionable means— point to the desire among the public for the bodies to have been donated consensually, for their presence there to be "true," albeit anonymous. In contrast, Bentham's statements in his will demonstrate his eagerness to have what was unique to him preserved and displayed so that he will always be remembered *as he was*. The preservation of his body was to tell his story as a kind of material life narrative for generations to come. His singularity as Bentham and as a Utilitarian philosopher is preserved in the display of his chair, clothes, and walking stick, and he has succeeded in making an icon of himself such that he will neither be put away nor forgotten.

The Auto-Icon is, in the end, both an archive of the details of Bentham's most intimate identity as well as a material autobiography. Bentham is eternally frozen in the pose he held when writing the work that would form the basis of the archive he would eventually donate to the University of London, later University College. Like Snow White, he sits in a sort of three-dimensional, visible crypt, within the archive that is the university, home to the actual archive of his life and letters housed at the Bentham Project. He has made a symbol of himself— icon — in his own image, which is, in fact, the task of autobiography that affords the author the opportunity to pick and choose in shaping a wholly new textual identity. This *mise-en-abyme* effect of the Auto-Icon, and its extensive posthumous travels and activities, help it to serve as a symbol of the museum or collection's capacity to construct and communicate identity. Although the university is not a museum *per se*, the display of the Auto-Icon has rendered it as a museum space: Bentham cites the desire for his display to educate the public about the human body, and he created a foundational text in order to create a rationale and context for the public as opposed to the private nature of his display. However, at the heart of this exhibition is a very intimate, private, and deep-seated desire, on the part of the participant, to surmount death, and on the part of the viewers to see what is wondrous, unseeable or unrepresentable. What cannot be represented here is the abstraction, the construct, the patched and fragmentary costume that is human identity itself, signified by Bentham's body on display.

And in a sense, though they remain anonymous, the subjects who comprise the collection known as Body Worlds are just as iconic as Bentham. It is their lack of defining characteristics and seemingly generic physicality that reminds the viewers of the wonder inherent in us all, wondrous precisely because our anatomy is invisible even to ourselves even as we live in and observe our own bodies every day. Viewing this collection is similar to the effect of looking at one's bones in an X-ray: one sees in a shocking way the core of what lies beneath the quotidian and the familiar. Like the work of poetic language, the exhibition's power lies in its defamiliarization of the everyday. The very privileged glimpse into the literal inner lives of the specimens becomes like reading someone else's diary or overhearing a private conversation: we know we shouldn't be there but we are.

The viewers of exhibitions such as Body Worlds and the Auto-Icon, too, achieve a certain immortal power in the presence of the dead. Our hope is that by looking within (using anatomy) or by looking without at a subject who is now an object, who cannot stare back or break or return our gaze, we the museumgoers will learn something about ourselves that we have been unable to see except through the projection onto an absolute and radically different, objectified other. This process of using the object to explain oneself to others or others to oneself, to reveal the face of a subject's inner life and identity, is the unique power of material life writing, a kind of personal narrative made possible by the anatomical collection.

Notes

1. Nicoll, "The Auto-Icon."
2. von Hagens, "The Anatomist's Hat."
3. Crimmins, *Bentham's Auto-Icon and Related Writings.*
4. Ibid., Introduction.
5. Ibid.
6. Ibid., xxv–xxxix.
7. Freud, "The Uncanny," 217–252.
8. Hoffmann, "The Sandman," 85–126.
9. Freud, *Totem and Taboo.*
10. Bentham, *Will*, 16.
11. *Oxford English Dictionary Online*, s.v. "Anatomy."
12. Anemone, "The Monsters of Peter the Great."
13. Gould, "Dutch Treat," 17.

14. Kujundzic, *The Returns of History*, 139–40.
15. Ibid., 143.
16. Stewart, *On Longing*, 152; Cardinal and Elsner, *The Cultures of Collecting*, 1.
17. Benjamin, "Unpacking My Library," 66.
18. Cardinal and Elsner, 179.
19. Stewart, 161.
20. Ibid., 156.
21. Ibid., 158.
22. Ibid., 162.
23. Ibid.
24. Benjamin, 67.
25. Baudrillard, "The System of Collecting," 7.
26. Bal, "Telling Objects," 98.

Bibliography

Anemone, Anthony. "The Monsters of Peter the Great." Home page of Anthony Anemone. Williamsburg, Va.: College of William and Mary. http://aaanem.people.wm.edu/MonstersPeter.pdf (accessed August 4, 2005).

Bal, Mieke. "Telling Objects." In Elsner and Cardinal, *The Cultures of Collecting*, 97–115.

Baudrillard, Jean. "The System of Collecting." In Elsner and Cardinal, *The Cultures of Collecting*, 7–24.

Benjamin, Walter. "Unpacking My Library." Trans. Harry Zohn. In *Illuminations*, edited by Hannah Arendt, 58–67. New York: Schocken Books, 1968.

Bentham, Jeremy. *Will*. In Crimmins, *Bentham's Auto-Icon*, 1–18.

Crimmins, James E., ed. *Bentham's Auto-Icon and Related Writings*. Bristol, England: Thoemmes, 2002.

Elsner, John, and Roger Cardinal, eds. *The Cultures of Collecting*. London: Reaktion Books, 1994.

Freud, Sigmund. *Totem and Taboo*. Trans. A. A. Brill. New York: Vintage Books, 1961.

_____. "The Uncanny." Trans. James Strachey. In Vol. 17 of *The Standard Edition of the Complete Psychological Works of Sigmund Freud*, edited by James Strachey and Anna Freud, 217–252. London: The Hogarth Press, 1974.

Gould, Stephen Jay. "Dutch Treat: Peter the Great and Frederik Ruysch." In *Finders Keepers*, edited by Rosamond Wolff Purcell and Stephen Jay Gould, 13–32. New York: W.W. Norton, 1992.

Hoffmann, E.T.A. "The Sandman." Trans. R. J. Hollingdale. In *Tales of Hoffmann*, 85–126. London: Penguin, 1982.

Kujundzic, Dragan. *The Returns of History*. Albany, New York: SUNY Press, 1997.

Nicoll, Irena. "The Auto-Icon." UCL Bentham Project, April 4, 2007. http://www.ucl.ac.uk/Bentham-Project/Faqs/auto_icon.htm (accessed September 28, 2007).

Stewart, Susan. *On Longing: Narratives of the Miniature, the Gigantic, the Souvenir, the Collection*. Durham, NC: Duke University Press, 1993.

von Hagens, Gunther. "The Anatomist's Hat." Gunther von Hagens' Body Worlds: The Original Exhibition of Real Human Bodies. Heidelberg: Institute for Plastination, 2006. http://www.bodyworlds.com/en/gunther_von_hagens/anatomist_hat.html (accessed February 10, 2007).

Section Three

The Resurrection of Excoriated Bodies

Affecting Bodies*

Natalie Loveless

(de)Monstrations

AUSCHWITZ. AUSCHWITZ. CHERNOBYL. ART MAUDIT. AUSCHWITZ. TITANIC. CHERNOBYL. HIROSHIMA. NAGASAKI. TERRORISM. SITUATIONISTS. LOS ALAMOS. NEUENGAMME. PIETA OF AVIGNON. GUERNICA. MASS MEDIA. REPRESENTATIVE. PRESENTATIVE. PITILESS. REPRESENTATION. NEW NEUROTIC REALISM. TABLOID. WALL OF THE FEDERATED COMMUNE OF SUICIDES. OVEREXPOSURE. RAVE PARTIES. BACKROOM. SELF-DESTRUCTION. PITY. PIETY. TITANIC. REPRESENTATION. PHOTO-FINISH. LIVE COVERAGE. CNN. INTERACTIVITY. MEPHISTOPHELES. AUTO-DA-FE. TERMINAL ART. AUSCHWITZ-BIRKENAU. PHOTO-FINISH. BLITZKRIEG. PORTRAITS OF THE FAYOUM. PHOTO-FINISHES. THE LOOK OF DEATH. PRESENTATION. ABSOLUTE PRESENT. REAL TIME. MONSTER. SUMMUM. BIOLOGY. GENETIC GENIUS. SCIENCE. BIOLOGY. DNA. HOMUNCULUS. AUSCHWITZ-BIRKENAU. PROCREATION. CREATION. HUMAN. TRANSHUMAN. INHUMAN. EXTREME ART. COUNTER CULTURE. COUNTER-NATURE. SNUFF LITERATURE. SNUFF MOVIES. SNUFF VIDEO. SNUFF DANCE. THANATOPHILIA. VIVA LA MUERTA. TERMINATOR. TERATOLOGICAL. NUREMBURG CODE. SCIENTIFIC. ARTISTIC.

The words above punctuate the pages of Paul Virilio's 2004 polemic *Art and Fear*. They appear in this order. They appear in capital letters. They typographically mark the field of the page, providing visual interventions in a sea of regulated text — text suggesting that artistic abstraction be read as representative of a sadistic impulse traceable directly to the violence of World War I, "when the first abstract canvases appeared and the human figure was literally

The following is based on site research and interviews in London (2002), Los Angeles (2004), and San Jose (2007) funded by the History of Consciousness department of the University of California, Santa Cruz, and the Institute for Humanities Research. I owe a debt of thanks, in particular, to Felicity Ruperti, Body Worlds press agent, for facilitating access, materials and stimulating conversation both in London and Los Angeles. For materials and aid in accessing the Bentham archives and Bentham's mummified head, as well as facilitating time with James Hales, Head of the Auto-Icon conservation team, I am grateful to Catherine Fuller of University College London's Bentham Project. Thanks also to Jacques Euphrasie, curatorial assistant at the Musée Fragonard, for access to and background information on Honoré Fragonard's anatomical specimens.

and figuratively blasted to bits."[1] Asserting that the destruction of the human form in representational abstraction exists in an intertwined and causal relation with the destruction of the human body, *Art and Fear* stages a face-off between what Virilio calls *demonstration* and its sadistic other, *monstration*, a practice that works "to outdo mere REPRESENTATION by offering *the very presence of the event.*"[2] One of the examples Virilio offers as a foil in this polemic is Gunther von Hagens' controversial anatomical exhibition Body Worlds.

Rather than engage in a general debate about representational practice (what it might mean for *monstration* to take over the mediating function of *representation*), Virilio's claims will be illustrated in this essay specifically through Body Worlds. The exhibition offers a particularly charged location from which to think through the stakes of Virilio's text, one that profits from reading the shifting exhibition codes that have accompanied Body Worlds in its move from Europe and Asia to North America. An examination of these shifting exhibition codes will also be used to ground us in a historical lineage of controversial post-mortem practice that finds a particular affinity in Jeremy Bentham's 1832 Auto-Icon. Ultimately, this essay considers Body Worlds as a project that, rather than simply contributing to debates surrounding techno-scientific excursions that cross a putative art/science divide (excursions that Virilio's text is explicitly positioned against), offers a particularly charged location from which to think through the stakes of affect and identification as pedagogical tools.

Before turning to my discussion of the exhibition proper, then, it is necessary to pause on these two terms, affect and identification. Though both of these terms have a place in common and philosophical discourse, my use of them in this essay stems from their use in psychoanalytic theory. Psychoanalytic theories of affect play a central role in understanding the role of trauma and memory in Freud, transference and the subject-Other relation in Lacan, and the promise of the analytic scene in both. For the purposes of this essay, I am primarily interested in Lacan's use of affect to assert that the commonplace of a binary split between emotion and intellect is not valid within the psychoanalytic field. This understanding, within which the affective cannot be understood as a separate realm opposed to the intellectual, provides a counterpoint to Virilian understandings of representational practice—a counterpoint through which affect emerges *within* identification. Identification, while generally understood as either a classificatory act (a relation of difference: *to identify*) or a constitutive act (a relation of similitude: *to identify oneself with*), in Freud's work, through his development of the Oedipus complex, comes to denote "the operation itself whereby the human subject is constituted."[3] To this operation, Lacan, in turn, adds a special emphasis on the role of the *image*—a role embroiled in complex negotiations of ambivalence and aggression.

Having laid out a gross terrain for affect and identification, let me move forward with the hope that, even without a full engagement with the psychoanalytic stakes of the terms, something of the enmeshed and constitutive aspects of these functions within a psychoanalytic theory of the subject will nonetheless point towards a conception of representational practice at odds with the Virilian model. Without attempting to fully account for the varied and complex histories of these terms, even within psychoanalysis itself, I would like to highlight the way that each of these terms, held separately and together, offers an interpretive framework for a reading of the particular pedagogy that I see at work — as a proposition, if not always a practice — in Body Worlds.

An anatomical exhibition that has amassed twenty-five million visitors in eleven countries since its unveiling in Japan in 1995 (at the 100th anniversary of the Japanese Association of Anatomists), Body Worlds uses a preservation technique called plastination, a process allowing preservation of the dead body in incredibly durable, life-like, and dramatically upright poses for a projected 1,000 years, to sculpturally mold dead bodies into statuary. The particular technology and labor invested in these specimens provide an image of upright integrity and life-force, and, rather than the disorder of decomposition, offer us an anatomical structure as accessible, clearly displayed, and stable. The plastinates, on the one hand, recall images from Renaissance anatomical and early twentieth-century art historical texts: from Andreas Vesalius, the La Specola waxes, and Frederik Ruysch's anatomical tableaux — one of which famously featured a fetus crying into a handkerchief made of placenta, surrounded by vascular systems as foliage — to early twentieth-century artworks of Naum Gabo, Umberto Boccioni, and Salvador Dalí. On the other, the plastinates mimic cultural caricatures through function and form (the baseball player hitting a home run; the chess player leaning over a chessboard, frozen mid-move).

Divided in its formal approaches, the exhibition offers, at one extreme, bodily parts displayed according to conventional codes of modern anatomical display (the juxtaposition of healthy lungs and tar-stained ones; cross sections producing two and three dimensional configurations; fetuses from between 17–33 weeks, including some with birth defects). At the other, it offers a series of what von Hagens calls "gestalt plastinates" — whole bodies articulated in whimsical, active, and sometimes fantastical poses. *The Runner* (a figure whose muscles fly backwards off of his bones, revealing different layers of muscle, both the outer and the lower muscles simultaneously) mimics Umberto Boccioni's 1913 Futurist sculpture *Unique Forms of Continuity in Space*. *The Drawer Man* (a plastinate cut into drawers that shows how neatly packed the various body structures really are) recalls Salvador Dali's 1936 Surrealist sculpture *Venus de Milo With Drawers*. *The Rearing Horse with Rider* — the *Rider* flayed and split twice longitudinally to reveal underlying musculature while coaxing a plastinated horse over a jump — pays homage to the eighteenth-century professor of anatomy Honoré Fragonard's *Horseman of the Apocalypse* (a preservation of a horse and rider that can be seen, along with Fragonard's other metal alloy preservations, in the Musée Fragonard at the Ecole Nationale Vétérinaire D'Alfort, Paris). Like von Hagens, Fragonard's aim was to trans*form* the body into a statue without trans*lation* into marble, plaster, or bronze. To produce the *Horseman* he injected a metal alloy into the specimen's arteries, and, after the remaining tissue dried out, the entire preparation was treated with lacquer. The resulting anatomical preparation offered a permanent display of individual anatomy that aimed to rectify the lack of cadavers available for live dissection. The technology available to Fragonard, however, left him with models that, while impressive in their own right, offered limited insight into the finer details of bodily systems.[4]

Von Hagens came up with the idea for exhibiting the plastinates in this way while using models and sections in lectures with students of anatomy at the University of Heidelberg. In his words, "people always went towards the intact bodies. They saw themselves in the specimens, and gave them names. That gave me the idea for my first exhibition ... I developed a kind of exploded view, moving body parts so that lay-people can put them back together in their minds."[5] This statement, in my reading, highlights the interrelatedness of the functions of affect and identification for von Hagens: the spectator is asked to identify with the exploded body, see in it the partition of her own animate body, and, through an identification with the similitude of the plastinate, to also engage in an affective relation with its former life as if invoked by the plastinate's formal articulation.

Exhibition Codes

The first Body Worlds exhibition that I attended was in 2002, at the Atlantis Gallery in the Old Truman Brewery, Brick Lane, London. Over the course of two afternoons I meandered, encountering bodies live and dead: the gestures of viewers moving in to follow the gestures of plastinates, frozen in the act of displaying their bodily systems like Vesalian engravings— although with more heroicism and less abject *vanitas.* While the organs were in cases, the gestalt plastinates were not. Unprotected by ropes or glass, they remained accessible, available. One could scrutinize as closely as one might wish, and even the occasional touch was often noticeable, if not encouraged (except, of course, in the case of the "touch-tours" led for the blind).* Whereas the conditions of display in North America have progressively made such interaction impossible — the 2007–2008 San Jose exhibition, for example, has the majority of its gestalt plastinates under plexiglass— the London exhibition offered the viewer-participant a complex and heterogeneous space. The strict linearity of the viewing narrative I found in the Los Angeles and San Jose exhibitions— where one is guided through the exhibition space by placards, lighting, and dividers— was

**See* Goodbye Body Worlds, *a pamphlet produced and distributed at the end of the UK run of Body Worlds, for a brief discussion of one of these tours.*

absent. While there *were* placards and lighting, the choreography of the space in London was looser, primarily delineated through installations of natural foliage. It even made use of the gallery's food court, set halfway through the exhibition space, where one could contemplate the bodies while sipping on tea from slightly above, but still very much within, the exhibition itself. This loose organization suggested, in its experimental ingenuity, that we each find our own way through the space. Due to its size, organization, and lack of dividing curtains, it seemed that we were constantly bumping into each other, as uncertain of how to navigate the space as how to read the objects.

I was regularly witness to fingers that, in pointing something out, moved in to touch — just to see — and I saw in this gesture two central aspects of the Body Worlds encounter: the entwining of the plastinates' formal spectacle with their didactic function. Bereft of biographical history — von Hagens insists on donor anonymity — the exhibition highlighted the individuality of these bodies at the level of structure. As the official plastination donation brochure put it, "as a rule we do not plastinate human specimens for private persons. We are discussing making an exemption for contemporary historical figures who wish to and should be plastinated in order to be kept in a mausoleum or crypt."[6] Divorced from the biographies that once articulated their flesh, and molded into caricatures that ostensibly took bodily function as their point of origin, the plastinates were both powerful to behold and difficult to read, and it was this difficulty that kept me transfixed. I was attracted to the way that Body Worlds worked to challenge the framing of disciplinary and institutional specialization, and saw it as a uniquely interdisciplinary project, one that asked its viewers to inhabit a de-familiarizing space both at the level of exhibition practice and pedagogical practice.

That said, from its first wave in Europe to its arrival, in 2004, on North American shores, there has been a noticeable shift in exhibition framing and language. While the exhibition originally suggested an uneasy collapse between conventional aesthetic and scientific codes, with Body Worlds' arrival in the United States, von Hagens has gone to great pains to represent the project as a primarily scientific endeavor, arguing that the exhibition should be read only as democratized medical knowledge production rather than as a form of transgressive aesthetics, a challenge to taboos surrounding the treatment of the dead, or a revolution in funerary practice. In keeping with this, Body Worlds in the United States has been framed squarely within a conventional "natural history" or "science museum" lens — a turn at odds with the exhibition's early history, one that insisted not only on raising questions of morality and taboo (ideological narratives articulating the post-life body as alternately utilitarian and sacred) but also on situating the exhibit in alternative, non-scientifically-coded venues (e.g., the Atlantis Gallery at the Old Truman Brewery in London, the old Mail Railway Station in East Berlin, an old abattoir in Brussels and the *Erotic Art Museum* in Hamburg).

These early exhibitions had, as their central strategy, a proposition regarding humor explicated during the 2002–2003 London exhibit. This Body Worlds featured a series of what von Hagens called his Mystical Composite Plastinates along with an article for the press entitled "Or Why it is ok to laugh at dead bodies." The article argued that the Mystical Composite Plastinates demonstrated anatomical features in a way that surpassed what was possible from naturalistic, or serious, poses, opening "a world of fantasy which allows us to accept our mortality easily."[7] One of these, a witch hanging from the ceiling on puppet strings, rode its own internal organs like a broom, its facial muscles flipped back from its skull to form a magician's cap. The *Winged Man*, for its part — sporting one of von Hagens' own signature black hats— offered back muscles fanned out like wings and a peace sign carved on its abdomen. The audio tour commented that this black hat functioned to "further close the gap between life and death [and] help break another taboo which blocks our mind [that] dead bodies should not be played with or laughed at. To play with them helps to deal with fear of the unknown."[8] Through an encounter facilitated by what von Hagens calls "non-serious" poses, plastination was thus proposed as a forum forcing "unavoidable discussions as to whether it is ethically defensible to turn a human body into a work of art."[9] In another interview he states: "What I do is not art, nor is it science. It reaches into artistry but the effect goes beyond education because emotions are involved."[10] Such discussions are precisely what emerged around Body Worlds in Europe.

From Cologne, Vienna, Basel, and Berlin to Hiroshima, Seoul, and Busan, claims against Body Worlds included that it denied the body's eternal rest, sensationalized rather than educated, and deliberately transgressed boundaries between morality, art, science, and commerce by treating the human body, as one journalist put it, "as something tantamount to a sculptor's clay."[11] During the run of the Berlin exhibition, a requiem mass was held to pray for the souls of the dead; promotion for the exhibition included placing plastinates in a local snack bar and sending a pregnant odalisque around the city, behind glass on the back of a bus, her belly open in the middle revealing an eight-month-old fetus. In Mannheim, Roman Catholic and Protestant Church leaders denounced the exhibit as a breach of human dignity and attempted to close the show, arguing that "the human body shouldn't be used for amusement."[12] Prime Minister Teufel himself went on television to officially denounce the exhibit, and local district prosecutors began the process of pressing criminal charges against museum officials (stating that it violated the sanctity of the human body for the purposes of commercial art, they almost succeeded in preventing the exhibit from opening). In Brussels, where Body Worlds was shown in an old abattoir with vaulted ceilings, Belgian physicians called for a boycott of the exhibition. And lest we think reception was one sided, while representatives of the Catholic Church in Cologne officially asked people to boycott the exhibit, in Oberhausen the head of the Catholic college himself invited Body Worlds into town, having learned that young women viewing the embryos were saying that they would never again consider abortion.[13]

The London exhibition—where von Hagens had *The Rearing Horse and Rider* installed at the *Hard Rock Café,* costumed for Halloween—had no less controversy than elsewhere in Europe. Only one year after the Hayward Gallery's sensational *Spectacular Bodies: The Art and Science of the Human Body from Leonardo to Now* (an exhibition that brought together more than 300 works of art, anatomy, and science from fifteen countries, making explicit connections between Renaissance and contemporary work), and with the likes of Damien Hirst, Rick Gibson, and Mark Quinn still in the limelight (all contemporary artists who made their names producing work, intended strictly for a contemporary art audience, out of biological and decaying materials), von Hagens was well positioned to encourage a challenging set of questions around the exhibition through press conferences, publicity stunts, and other attempts to upset divides between scientifically, commercially, and culturally-coded space.

On March 26, 2002, Geoffrey Lee, 51, entered the London exhibit and smashed the plastinate called *The Organ Man* (a man parted down the middle, holding his liver extended in one hand) with a claw hammer. "I never believed in such a thing as evil until I saw human bodies had been turned into commercial objects," he told the press.[14] Two days earlier, Martin Wyness had thrown a blanket over the *Reclining Pregnant Woman* plastinate and spilled paint on the floor as an act of protest (Wyness later went on to protest outside the exhibition, an action von Hagens acknowledged by joining him for a public debate).[15] Repeating the drama of Berlin, in London several members of parliament tried to stop the show from opening, claiming not only that it undermined human dignity but also that it went against Britain's 1832 Anatomy Act. Her Majesty's Inspector of Anatomy, Jeremy Metters, however, was forced to announce, only days before its scheduled opening, that there were no laws in place that could ban Body Worlds.

These protests in England were, in part, a result of tensions running high in the wake of the Alder Hey hospital organ retention scandal, in which it came to light that between 1988 and 1995 organs were removed, retained, and disposed of without family consent (that the organs removed included those of children became the heart of the debate), and in this wake, many called on the Anatomy Act for vengeance. Passed in 1832 to regulate traffic in body parts and protect against the practice of body snatching (as exemplified most famously in the Burke/Hare grave robbing trials), the Anatomy Act could not be held to apply to Body Worlds because it specifically covers dissection and medical traffic, not display—leading to the plastinates, in this case, being categorized alongside the Egyptian mummies and bog people displayed in the British Museum. Though nothing could be legally done to prevent its opening, controversy surrounding the exhibition led to the production of legislature working to ban the return of Body Worlds to England. In protest against this new code — the United Kingdom Department of Health's 2002 "Draft Code Of Practice On The Import And Export Of Human Body Parts" — von Hagens extended the stay of the exhibition (with posters stamped "Assigned for Deportation") and staged a televised anatomical theater, the first public dissection in the United Kingdom in just under 200 years (1830), modeled on Vesalius' sixteenth-century theatre of anatomy in Brussels.

In the move from Europe to North America, most elements of transgressive laughter, protest agitation, and moralistic discomfort have been suppressed in favor of a normative didacticism that suggests not only how to live a healthy life — as evidenced in the central place of the smoker's lung in exhibition practice and literature — but also, at the San Jose exhibition, in how to live a gendered life: one of the wall placards states that "on average, women are more empathetic, expressive, articulate, and diplomatic, and better able to sense what others feel ... [while] men may be better at focusing on a specific task, orienting themselves spatially, or solving complex math or physics problems."[16] Not to suggest that this normative didacticism was absent from earlier incarnations of Body Worlds (Indeed, this question of gendered representation has dogged the exhibition at least since London, where Amanda Wilson stripped down to her underwear and jumped on *The Rearing Horse* to protest that female plastinates were, for the most part, relegated to the display of the reproductive system. Von Hagens has responded to these initial criticisms by including more visibly female plastinates, though there are as yet no female musclemen, baseball, or chess players on display, instead taking the form of plastinates like *The Angel* and *The Yoga Lady*). Rather, I would like to suggest that, despite the normative gender and anthropocentric basis of the exhibition, the alternative venues and controversial practices that marked Body Worlds in Europe displaced the act of plastinate viewing from the suturing lens of *data* to a foregrounding of the *scene* of encounter. This is not to say that the scene of encounter (and the reality effect that is attached to this) has been completely effaced from current incarnations of Body Worlds or the exhibition literature surrounding it. Rather, it is to assert the important role that alternatively coded venues had, in pre–United States incarnations of the exhibition, in displacing assumed codes of legibility and function. Since its arrival in the United States, site, context, and framing narratives emerging from the von Hagens team have insistently policed references to art, transgressive aesthetics, or the creative expansions we could call sculptural post-life practice. (Felicity Ruperti, in the "Genetics & Plastination" section of *Goodbye Body Worlds,* also discusses von Hagens' search for a terminally ill donor "willing to discuss with scientists how his body will be dissected and restructured as a 'super-human' after his death" — a project aiming to "visualize what a genetically restructured human might look like" and that would be televised and thus end anonymity as a hard and fast rule of plastination.[17]) And, in all of this, instead of the issue simply lying in the turning of a "someone" into an "object," what has been at stake is the *kind* of object. An object dedicated to popular anatomical knowledge production redeems itself (or its maker); one that adds to this a disruption of conventional taboos and rituals surrounding the material we call a dead body incites riot.

Social Utility

It is not only anatomy that von Hagens brings back to life by fleshing out Vesalian imagery and Renaissance waxworks, but also a dream long gone proposed by the Enlightenment philosopher Jeremy Bentham in his last, and unpublished, 1831–32 essay, "Auto-Icon or Farther Uses of the Dead for the Living." In correlation with Bentham's last will and testament — in which he requests that his body be publicly dissected before being made into an Auto-Icon — this essay suggests a utilitarian program for the materiality of the body after death: one in which, "with the consent of parties concerned, the mass of matter which death has created, be disposed with a view to the felicity of mankind."[18] In his essay, Bentham defines Auto-Icon as the preparation, preservation, and display of the dead body for social use: "*Auto-Icon* will be soon understood for a man who is his own image."[19] While there are significant differences between Bentham's and von Hagens' visions, they profit by being read together as both controversial and conservative propositions with regards to sculptural post-life practice.

It is said that for the ten years leading up to his death, Jeremy Bentham carried a set of glass eyes with him everywhere he went.[20] These eyes were to be *his* eyes, inserted into his preserved head to figure his post-mortem sight for eternity. Taking seriously the question of post-mortem utility, and eerily pre-figuring Body Worlds, Bentham set the stage for this post-humous self in "Auto-Icon or Farther Uses of the Dead for the Living," describing a future in which "entire museums of Auto-Icons would be formed."[21] This vision is what constitutes the "Farther" in Bentham's title (the "nearer" being the promotion of anatomical dissection, the naturalization and decriminalization of which Bentham was active in promoting). Bentham's title refers to Southwood Smith's article in the *Westminster Review* in 1824, "Use of the Dead to the Living." The article promotes the "immeasurable utility" of anatomical dissection in the training of surgeons, and suggests that the bodies of the unclaimed dead be made available for dissectional practice in order to prevent that practice being done on the bodies of the living (with the understanding that, in either case, it is the poor who will suffer, the choice here being between the living and the dead poor). This is, in fact, what, in 1832, the Anatomy Act legislated, leading to the collapse of the body snatching trade. The oration over Bentham's corpse and public dissection, performed by Southwood Smith eight years later while the Anatomy Act was in parliament, became a public call to add to his initial formulation a strong plea for individual bequests. As such, Bentham's body, during the oration, became exemplar.[22]

"Auto-Icon or Farther Uses of the Dead for the Living" suggests, in addition to the use-value of a corpse for anatomical dissection, a secondary use — "a permanent, — say a conservative, or statuary."[23] Cross-temporally echoing both Fragonard and von Hagens, Bentham champions the wide-scale use of this as a practice of *auto-thanatography*, telling his reader: "If a country gentleman had rows of trees leading to his dwelling, the Auto-Icons of his family might alternate with the trees."[24] Further, in suggesting that statuary and coats of arms be replaced by the very bodies of those meant to be represented, Bentham submits that, with his Auto-Icon, "a spick and span new subject-matter of property (would be) brought for the first time into existence," one that, in its authenticity, enshrines a post-mortem future, diminishing the horror and increasing the utility of death. While von Hagens insists, in the production of a plastinate, on erasing personality and preserving only structure, Bentham's proposal is grounded in the utility of the Auto-Icon as indexically linked to the person who preceded it: "As regards anatomical purposes, one body serves for instructing nearly as well as another; but the head of each individual is peculiar to him, and, when properly preserved, is better than a statue."[25]

Though Bentham's remarkable essay addresses a wide range of detailed, sometimes humorous, and highly imaginative uses for the permanently preserved dead body (including economic, anatomical, phrenological, moral, genealogical, commemorative, and architectural), it is firmly committed to moral instruction and offers itself as a limit-case exercise in utilitarian philosophy. Most interesting of Bentham's uses in this regard are those "Theatrical or Dramatic" that he describes under the title "Dialogues of the Dead":

> By means of strings or wires, by persons under the stage, or if the Auto-Icon were clothed in a robe, by a boy stationed within, and hidden by the robe, the eyes being already made in imitation of his, the eyelids might be made to move; and in so far as needful or conducive to keeping up the illusion, the hands and feet, one, more, or all. As to voice, by well-known contrivances, it may, without difficulty, be made to appear to proceed from the vocal organs of the figure; the body, if necessary, might by obvious contrivances, be made to appear to breathe.[26]

Founded on affective identification, this fantasy offers an alternative in which a king would be "literally imperishable," in which a philosopher could be forever visited, with his "unfinished manuscripts, lodged in an appropriate case of shelves."[27] As with the plastinates, such Auto-Icons would not only shed light on the archaic nature of taboos surrounding the treatment of the dead, bringing attention to the inefficiency of funerary practice, but would also function as exemplars for social behavior: how one acted in life would determine the treatment of one's Auto-Icon. Suggesting a temple of honor and temple of fame, into which the Auto-Icons of the elite would be placed according to their practices during life, Bentham tells us: "If injury had been done to the reputation of an Auto-Icon placed in the temple of dishonour, public opinion might redress the wrong — might correct its own mistakes— might transfer the sufferer to the temple of honour."[28] The sufferer, here, *is the Auto-Icon itself*, and, as in the Theatrical Uses above, Bentham's proposal depends on a slippage between the affective identification of the viewer with the Auto-Icon as both *same* and *different*— different from the viewer at the level of personality and identical at the level of a possible immortal and corporeal social utility.

That is, encounters with Auto-Icons of the virtuous function, for Bentham, as moral exemplars: by gazing on the faces of the virtuous or the fallen, in their permanent status as Auto-Icon, a kind of secular punitive hereafter is born in which one "must anticipate the judgment of his fellow men. If the judgment is unfavourable, he may be exposed with his head reversed."[29] Central to this form of moral instruction is an aesthetically preserved post-life body that is put to use in service of those left behind — a service that relies on a visual encounter that, through the relay of affective identification, moves the viewer to reflect on her choices in life. This moral imperative is one that is equally operative in Body Worlds.

Body Worlds, like Bentham's world of the Auto-Icons, offers a promise of socially useful immortality. And though, in Body Worlds, one's immortality is, at present, anonymous, one is, nonetheless, encouraged to *figure* one's personal physical existence beyond death. This is achieved through active solicitation for the Institute for Plastination's body donation program, including large scale Xeroxes presenting the handwritten testimony of chosen body donors, advertisements for yearly body-donor meetings including an expense paid trip to "Plastination City" in Dalian, China, as well as a prominent memorial, usually placed at the exit of the exhibition, to the body donors. The donation form, as well, ends with a list of reasons you can check off, should you be unsure, for choosing to donate your body for plastination. In order they are:

> I would like to donate my body to a good cause. / I am fascinated by the thought of being preserved forever for posterity. / I find the thought of conventional funerals, i.e. being cremated or buried, unpleasant. As there is no need for a burial service with plastination, my relatives will not have to worry about tending my grave. / I have no relatives to tend my grave once I am gone. / I wish to save on the costs of a funeral. / I am excited about the potential of plastination. / I am fascinated by the public exhibitions.

The form also includes a blank space for donors to articulate in their own words their reasons for becoming plastinates, examples of which are then published and distributed in the body donor packet. Take the following as examples: "I think about the idea that my human body will continue to teach. This is a kind of consolation";[30] "I just like the idea of cross-sections of my body traveling around. It's got to be more fun than being buried. And it's cheaper!" Articulating a fantasmatic attachment to the materiality of the body after death, these statements replace an image of bodily decay and invisibility with that of visibility and social use, on the one hand and a visibility *of* social use, on the other, that is grounded in affective identification. I would like to suggest that it is precisely on this point that von Hagens' pedagogical program, at its most interesting, turns. Calling the exhibition alternatively "event anatomy," "emotional anatomy," and "anatomy art," von Hagens insists that "We are changing the way we look into the future, I will survive in my body with a special purpose: the purpose of education ... I offer a death lived by anticipation, a death that crosses over into the sphere of life." [31]

Rejecting any attention to the *narrative* of individual lives, with its possible attendant sentimentality, von Hagens directs attention to the structural particularities of a body that, while it might be sensationally crafted, is nonetheless being framed primarily as an educational tool: "In a democratized anatomy, it is my task to lure the audience with the unfamiliar pose and to tame their sensationalism; to transfer it into a real interest in the body."[32] In service of this, in one article, von Hagens suggests that "the specimen acts as an optical bridge to his own body—he realizes who he is and what he's made of. This way he's emotionally touched."[33] In another article, the specimens function to "build bridges back to your own body."[34]

As a viewer I know that the plastinated body is real, and I engage in a fantasy of equivalence: *this is me, I am it*, or: *this could be me, I could be it*. Where anatomical models glide over differences between bodies in service of a world of anatomical averages, the imagined authenticity of a plastinate permits learning beyond what a mere model can provide. Linking the plastinates to Renaissance masterworks, von Hagens insists: "we have the need to see the real specimens. When people go into museums, they like to see the great masters, not reproductions. A layman has a thirst for authenticity, especially in a media oriented society where everything is a representation, not original."[35] For von Hagens, that learning can be located in the affective charge of an encounter with the plastinated body of one's posited material counterpart. It is this charge that articulates an uncanny pedagogical scene, grounding the value of his pedagogical program in affect and offering a focus in which the *scene* of encounter outweighs the *data* transmitted. That is, the exhibition has (whether foregrounded, as in London, or not, as in Los Angeles) a pedagogical commitment underpinning it, one attuned to the productive possibility of *affective* engagement—a commitment that is supported by both formal choices within the exhibition as well as in the rhetoric surrounding it: from strategic choices regarding where skin remains on otherwise flayed bodies—eyebrows, noses, nipples, navels, and genitalia—to exhibition literature, placards, and audio tours.

(re)Presentation

It might be interesting to read the Body Worlds narrative I have just offered as depicting a movement from the framework of radical democracy—a radical democracy that, to echo Chantal Mouffe "would be suspicious of any attempt to impose a univocal model of democratic discussion"—to that of a liberal democracy.[36] As a project in radical democracy, in its early incarnations Body Worlds presented us with the volatility of laughter, the championing of dissent, and the courting of protest action as central exhibition practices—ones that made sense of von Hagens' claim to encourage individuals in questioning their relationship to their bodies, to mortality, and to the received conventions of funerary practice. "They tell me that I market the human by, that I commercialize it. My answer is, there's nothing wrong with getting money from people who want to see the exhibition because, of course, there is a lot of money involved in putting it up. And when people buy specimens from me, they are paying for the plastination process, not for the specimen, because the specimen is donated. And I see no difference for an undertaker, who, for putting the body in the earth, gets money from the relatives at the worst point of their lives, when they've lost a loved one. I get money from the people who want to see the show, and I get money from the universities who want the specimens for medical teachings. I have no problem with that."[37] As a project in liberal democracy, Body Worlds has now turned to the fantasy of the citizen subject, to the effective civics lesson—a value located in the firmness, rigidity, coherence, and strength of the objects. It is the materiality of the plastinates, rather than simply their formal composition or context, that remains the heart of complaints like that of Virilio, with which we began.

In Virilio's reading, Body Worlds, by leaning into a history of pre-modern anatomical codes and conventions of visual display rather than the naturalized neutrality of modern medical display, imposes "its own terrorist voyeurism: that of death, live."[38] Echoing statements by von Hagens himself ("I want to bring the life back to anatomy. I am making the dead lifeful again. This exhibition is a place where the dead and the living mix"[39]) but towards a different end, for Virilio, the plastinate, as flayed and displayed corpse, paves "the way for actual torturing of the viewer."[40] Failing to permit heterogeneous affective responses, this statement dictates the terms of the viewing encounter, effectively closing down semiotic-affective possibility with the war cry of visual pollution. Recalling Medieval threats of visual contamination, and their attendant representational logics, for Virilio, plastinates *impose* themselves upon the viewer, bypassing the modifier of interpretation to torture us with the reality-effect of dissected and displayed human bodies. In this, Virilio enforces what he calls the *presentative* (*monstrative*) and *representative* (*demonstrative*) as distinct and oppositional structures, and it is on this opposition that his argument rests. Representation, in *Art and Fear*, functions as a liberatory process that is held in opposition to the collapse that ensues when, within Virilio's terms, we are offered something said to function as the thing-itself.

Virilio's text has remained useful to me as a particularly potent articulation of the need to decry not only Body Worlds, but all practices that pollute the boundaries of naturalist representation — whether Modernist abstraction, performance and action art, or bio-art experimentation. It reminds me of the degree to which I am invested in articulating a field surrounding Body Worlds, one that engages Body Worlds' assertion that affect is necessarily embedded within the putatively intellectual encounter, as well as one that engages the uncanny collapse this effects between conventional aesthetic, scientific, and pedagogical codes. Body Worlds, in its attempt to insert itself as a legitimate post-life practice, asks us to question our relationship to our own materiality, to the ecology in which we live, and to the cultural structures that legislate our actions. Ultimately it is the reading of all of these narratives together, a reading for a *conflicting plurality* of Body Worlds (from the radical to the liberal, the conservative to the avant-garde, the Benthamian to the Virilian), that offers the most potent terrain from which to think through the stakes of affect and identification as pedagogical tools— tools that not only benefit from the championing of dissent, but that are structurally organized around dissenting bodies, bodies that dissent and bodies in dissent.[41]

Notes

1. Virilio, *Art and Fear*, 29.
2. Ibid., 40.
3. Laplanche and Pontalis, *The Language of Psychoanalysis*, 205.
4. Simon, "The Theater of Anatomy," 69–76.
5. MacKenzie, "A Body of Knowledge," 22.
6. Institute for Plastination, *Body Donation Consent Form*, 20.
7. Rathgeb, "Or why it is ok," 2.
8. Ruperti, *Body Worlds Digital Audio Tour.*
9. Rathgeb, 2.
10. O'Rorke, "Skinless Wonders...," 22.
11. Andrews, "German Show Uses Corpses," 45.
12. Leth, "Museum's Body of Work," 11.
13. MacKenzie, 24.
14. Holmes, "Remains of the Day," 78.
15. Jones, "I Disagree Publicly," 6.
16. Ibid.
17. Ruperti, "Body Worlds and Organ Retention," 12.
18. Bentham, "Auto-Icon," 2.
19. Ibid.
20. Marmoy, "The 'Auto-Icon,'" 80–1.
21. Bentham, 6.
22. Richardson and Hurwitz, "Jeremy Bentham's Self Image," 195–198.
23. Bentham, 2.
24. Ibid., 3.
25. Ibid., 2.
26. Ibid., 10.
27. Ibid., 8.
28. Ibid., 6.
29. Ibid., 7.
30. Jeffries, "The Naked and the Dead," 33.
31. Institute for Plastination, 22.

32. Jenkins, "The Exhibitionists," 33.
33. Knipfel, "Body Art," 33.
34. Andrews, 33.
35. Knipfel, 33.
36. Mouffe, "Which Ethics," 94.
37. Knipfel, 22.
38. Virilio, 43.
39. Jeffries, 22.
40. Virilio, 37.

Bibliography

Andrews, Edmund L. "German Show Uses Corpses to Bring Alive the Body's Wonders." *International Herald Tribune*, January 8, 1998.

Bentham, Jeremy. "Auto-Icon; or Farther Uses of the Dead to the Living. A Fragment." Printed, not published c. 1842. Bentham Papers. University College, London, edited by John Hill Burton.

Evans, Dylan. *An Introductory Dictionary of Lacanian Psychoanalysis.* London: Routledge, 1996.

Finn, Peter. "Art of Controversy: Anatomy as Entertainment." *Herald Tribune International*, June 3, 2001.

Holmes, Pernilla. "Remains of the Day." *ARTnews*, May 2002.

Institute for Plastination. Body Donation for Plastination: Consent Form. First Edition, Heidelberg: May 2000.

Jeffries, Stuart. "The Naked and the Dead." *The Guardian*, March 19, 2002.

Jenkins, David. "The Exhibitionists." *The Sunday Telegraph Magazine*, November 10, 2002.

Jones, Toby. "I Disagree Publicly." In *Goodbye Body Worlds*. (2002): 6.

Knipfel, Jim. "Body Art: What a Piece of Work." *The New York Press*, July 1–7, 1998.

Laplanche, Jean, and J. B. Pontalis. *The Language of Psychoanalysis.* Translated by The Hogarth Press LTD. New York: W.W. Norton, 1974.

Leth, Frank. "Museum's Body of Work." *The Stars and Stripes*, January 13, 1998.

MacKenzie, Debora. "A Body of Knowledge." *New Scientist*, March 23, 2002.

Marmoy, C.F.A. "The 'Auto-Icon' of Jeremy Bentham." *Medical History* 2 (1958): 77–86.

Mouffe, Chantal. "Which Ethics for Democracy?" In *The Turn to Ethics*, edited by Marjorie Garber, Beatrice Hanssen, and Rebecca L.Walkowitz, 85–94. New York: Routledge, 2000.

O'Rorke, Imogen. "Skinless Wonders..." *The Observer*, May 20, 2001.

Rathgeb, Stephan. *Or, Why it is ok to laugh about dead bodies* (exhibition press kit). London: Arts and Sciences Ltd., 2002.

Richardson, Ruth, and Brian Hurwitz. "Jeremy Bentham's Self Image: an exemplary bequest for dissection." *British Medical Journal* 295 (1987): 195–198.

Ruperti, Felicity. *Body Worlds Digital Audio Tour.* Heidelberg: Institute for Plastination, 2002.

_____. "Body Worlds and Organ Retention Scandals." In *Goodbye Body Worlds* (2002): 8.

Simon, Jonathan. "The Theater of Anatomy: The Anatomical Preparations of Honoré Fragonard." *Eighteenth-Century Studies* 36, no. 1 (2002): 63–79.

Virilio, Paul. *Art and Fear.* London: Continuum Press, 2003.

Worrying About Democratic Values: Body Worlds in German Context, 1996–2004*

Peter M. McIsaac

Since Body Worlds' debut, one of the main justifications for the value of the show's controversial exhibition strategies has turned on the claim that Body Worlds democratizes medical knowledge. Though situated on one level as a critique of the way that medical science has developed and legitimated itself as a discipline since the early nineteenth century, the relationship of democracy and knowledge production has figured in other crucial ways with respect to the exhibition, both as an enabling condition and decisive factor in its reception during its first eight years. At this formative time, the German origins of the exhibition intersected with the fact that from 1996–2004, it showed primarily in Germany, where it attracted some 5.9 million paying visitors. Together, the various resonances and dissonances of these years have worked to mark the show, leading to its particular form and the peculiar success it has enjoyed precisely as a function of its highly controversial nature.

In this essay, I will be arguing that a distinguishing feature of the German engagement with the exhibition resides in the way it inflects German debates on the promotion of democratic values. My perspective is one that will regard democracy from a number of vantage points. Though affirmative notions of democracy, as they are shaped by the constitution and shaped by institutions, the media, and citizens, demand consideration, democracy in the post-1945 German tradition is also charged in obvious and not-so-obvious ways with maintaining distinctions vis-à-vis Germany's Nazi and GDR pasts. As I will show, consideration of these periods reveals democratic discourses to be an implicit, and implicitly disavowed, "other" of today's Germany. Thus, while the

**This essay is excerpted from a longer article in* Museums and Difference, *ed. Daniel J. Sherman. Bloomington: Indiana University Press, 2007, 181–194, 199–202. It is reprinted here with the kind permission of Indiana University Press.*

respective abuses of the Nazi and GDR eras often serve as potent reminders of the consequences of German authoritarianism, they figure only rarely in German discussions of Body Worlds. When the Nazi era, and the Holocaust more specifically, has surfaced, it has provoked disproportionate reactions from von Hagens and his organization, confirming the sense that the Holocaust and other authoritarian associations remain salient, if underdiscussed, issues with regard to the exhibition. While conducting debates on democracy as a proxy for thornier issues obtains in part due to the general tendencies at work in German culture and society, the exhibition's particular display and promotion strategies do more than exemplify these tendencies: they intervene in these discourses and set the terms of debate at a level that becomes perceptible using a museological analytical framework.

An important dimension to Body Worlds is how it stages its own relationship to critical voices and the debates about it. Practically every aspect of the show's construction and promotion has seemed predicated on presenting itself as a mode and model of enlightened liberation and tolerance. On one level, a vision of von Hagens himself is at stake in the exhibition, through which he presents himself as a guarantor of these values. Whereas the express desire to break taboos and crush superstitions is a part of this vision, another dimension couches plastination in terms of an at times personal struggle for democratic freedoms. Originally a citizen of the German Democratic Republic (GDR), von Hagens spent twenty-one months in an East German prison in the late 1960s for an attempt to escape to the West that floundered in part because von Hagens was unwilling to take all the risks that might have promised escape. Von Hagens was later one of the many East German prisoners in those years released to the West in exchange for financial compensation from the West German government. Arriving in West Germany in 1970, von Hagens resolved to make the most of his newly won freedoms and pursue his ideas and dreams to the bitter end. He recalls in the exhibition catalog that his thoughts on the day he invented plastination in 1977 centered on these political notions. In contrast to his wavering about whether to flee the GDR, in the West he embraced a "clear decision this time in favor of uncertainty, which is always an opportunity at the same time."[1] In his narrative, von Hagens stylizes himself as someone who has felt the struggle for democracy on his own body and as someone whose integrity deserves to be trusted. Specifically commenting on the relationship of von Hagens' biography to the significance of the show, Bazon Brock, a prominent German art historian, validates the exhibition's mythmaking on the basis of its maker's positive qualities. As Brock notes, "authenticity is established when one goes to jail for one's convictions."[2]

Authenticity in Brock's sense means refusing to hide in the face of controversy or danger, and it is through figures of brave self-exposure and openness that von Hagens' refusal intersects with the logic of Body Worlds as exhibition, on the one hand, and certain tropes of postwar German democracy, on the other. With respect to the former, von Hagens' endorsement of plastination's "uncertainty" inflects the technique as an embrace of knowledge that obligates its discoverer to expose his knowledge and the resulting burdens. Von Hagens would betray more than himself were he to suppress either the display of authenticated bodies or the slings and arrows of criticism that display has provoked: the presentation of critiques in the open visitor guest book is thus integral to a Body Worlds exhibition predicated on authenticity. In turn, the notions of self-exposure and openness resonate with long-established, postwar tropes of German democracy, signaling the workings of a government committed to antiauthoritarian principles, particularly in architecture and art exhibitions. The glass dome of Norman (Lord) Foster's renovated Reichstag that permits ordinary citizens to stand above and "oversee" the parliamentary chamber of the Bundestag is only the latest iteration of tropes of transparency and citizen empowerment found around the world in the anti-monumental, modernist glass and steel facades of West German government buildings.[3] In these buildings, acts of exhibition complement these architectural tropes, with over thirty-three million dollars having been set aside to procure and commission art by Sigmar Polke, Gerhard Richter, Hanne Darboven, Jenny Holzer, and Christian Boltanski.[4] Some artists, among them Hans Haacke and Thomas Schütte, refused commissions, seeing in them an attempt to co-opt art for political purposes and a papering over of the decades in which the German state had manipulated art exhibitions for ideological purposes, from the interventions of Emperor Wilhelm II around 1900 and the Nazis' degenerate art exhibits in the 1930s to the promotion of artistic programs by both East and West in the Cold War.[5] Though the exhibition of art in state buildings cannot be seen as either strictly perpetuating or rupturing these traditions, the government intends its embrace of contemporary art to represent its commitment to democratic openness, tolerance, and critique of authority. For the German government as for von Hagens, authenticity is won by ostentatious self-exposure.

Because plastination is thoroughly intertwined with von Hagens' story of integrity and authenticity, his narrative also locates Body Worlds subtly but squarely within the parameters of "democratic science," giving the impression that von Hagens implicitly disavows certain traditions. As Thomas Assheuer noted in the weekly *Die Zeit*, "If one did not know [that von Hagens had been imprisoned in the GDR for attempting to escape], one might believe that his necromania had come from the vaults of East German socialist science. It was notorious for its will to knowledge and for its slavish obedience [Kadavergehorsam, lit. "obedience until death"] with respect to progress. It resolutely believed that humanity could be delivered from myths and the fickleness of nature as well."[6] One only needs to recall the medical abuse of GDR athletes in the name of attaining enlightened socialism to have a sense of the destructive forms this science could take. Since von Hagens began studying medicine as a member of the Socialist party in Jena three years before his decision to leave the GDR, his current project may not be as divorced from socialist science as von Hagens' personal narrative or Assheuer's analysis implies. True, von Hagens claims that his experience of party hypocrisy disabused him of any desire to remain in the GDR.[7] Yet he has set up massive plastination facilities in Kyrgyzstan and China, which, as globalized business ventures, offer him lower operating costs and less stringent regulations than in Europe, in addition to connections he can exploit since he grew up in an East Bloc country. As he explained in an interview with Didi Tatlow, "I grew up in East Germany, and I really feel very at home in China.... It almost reminds me of my childhood. Little things, like the cups they use in hospitals. Also, I know how to deal with Communist Party secretaries. I know I'll get someplace with them."[8] Recent evidence that von Hagens' Dalian facility had accepted the bodies of executed Chinese prisoners casts doubt on the notion that Body Worlds can be readily dissociated from socialist science. Indeed, Assheuer's point about the abuses of enlightenment progress underscores the fact that the relationship between science and the attainment of democratic freedoms and protections is not automatic, particularly in light of twentieth-century German history. Before the GDR, the Nazis also sought to "enlighten" the populace about good bodily health (always linked to notions of racial hygiene) by using transparent, gendered anatomical models that traveled within Germany and abroad; they also carried out horrendous medical experiments on victims of the Holocaust in the name of "science."[9] Though real and meaningful distinctions might well be made between science in von Hagens' program and the respective GDR and Nazi eras, von Hagens' concepts of enlightenment, progress, and democratization might have been more extensively substantiated and argued in the exhibition.

These notions of enlightened tolerance are also inscribed in the exhibition by implementing, with modifications, the traditional technique of the visitor guestbook. Both in the exhibition and on the web, a whole spectrum of visitor comments is available for view. The over 70,000 comments visitors have made indicate that exhibition-goers feel a need to communicate with show organizers and other visitors.[10] However, the large numbers also mean that what can be viewed at any one time is a selection. Prior to 2003, Body Worlds seemed to go out of its way to select the widest possible range of opinion for placement on the web, with the show openly displaying passionate criticisms of its undertakings. Until recently it was possible to read a variety of strongly voiced opinions, including Christian Wetter's vehement attacks on "the pseudoscience of the show" and a German woman's wondering why von Hagens "has not yet been prosecuted."[11] Strongly held views in favor of the shows were also presented. But as evidence began to surface that von Hagens was receiving unwillingly donated bodies from Russian, and later Chinese, prisons, critics damningly referred to him as "Dr. Frankenstein" or a modern-day Josef Mengele. Changing his tactics, von Hagens began to lash out at critics who associated the show with Nazism. In light of revelations in February 2005 that von

Hagens' father had served with the SS in Poland, personal as well as social taboo zones may have been converging. A similar point can be made about comments made by von Hagens' wife and exhibition co-organizer, Dr. Angelina Whalley, in an article in Singapore's *Life!* magazine (also posted on the Body Worlds web site): "'If it's Dr. Frankenstein, it's okay,' says Dr. Whalley, 43, the medically-trained director of the Institute for Plastination in Heidelberg. But she takes offence at comparisons to Nazi concentration-camp doctor Josef Mengele, who conducted grisly experiments on inmates. 'We are always open to any criticism and we accept that. But if it's pure maliciousness, of course, it's something we cannot accept.'"[12] Reserving the right to reject outright critiques it deems as "malicious," Body Worlds' presentation of opposing views has diminished, with positive assessments and general references to the show's controversy replacing explicit criticisms. Nevertheless, visitors are still prompted to share their views before reading those of others, giving the impression that the show's promoters continue to tolerate, if not welcome, opposing viewpoints. What can be seen as a commitment to the democratic value of tolerance on one level welcomes controversy on another, translating, as several commentators have noted, into increased visitor numbers and better business.[13] The resonance of democratic concerns in the German reception of Body Worlds tracks the shapes of larger German debates. For both supporters and detractors of the show, nothing less than full commitment to democratic values proclaimed in the German Constitution has been at stake in this exhibition. This is the case even when the bulk of the arguments are presented in discourses other than law, such as art history, medicine, history, or philosophy. Supporters point to the fundamental right of well-informed, mature citizens to decide what happens to their bodies before and after death. In this view, a democratic state should on principle not prevent von Hagens from applying his methods to their corpses since there is no public health risk after plastination. Proponents also stress an intention to equip ordinary citizens with previously privileged medical knowledge and thus better control of their own bodies. Detractors maintain that the right to proper burial (*Bestattungszwang*) and the protection of human dignity (*Würde des Menschen*), cornerstones of the post–Holocaust Constitution, are being violated for base, money-making spectacle.

Recent German scholarship underscores that the functionality of democracy is principally at stake in the German evaluation of this exhibition. As Franz Josef Wetz and Brigitte Tag, the editors of the first scholarly essay collection on Body Worlds, explain, their volume is motivated by their perception that the debate on the show had failed to proceed in an ideally democratic manner. For them, Germany has still not moved far enough away from certain "painful experiences of history," a reference to both the Holocaust and, to a lesser extent, the authoritarian GDR.[14] Among other forms of repression, these regimes of course prohibited open debate and critical opposition. In Wetz and Tag's anti-authoritarian model, democracy implicitly disavows those eras whenever it is invoked, precisely because it entails a constant renewal of traditions under the scrutiny of reason and toleration of opposing views. Like Wetz and Tag, I wish to regard practices and discourses related to democracy in terms of how they enforce differentiations between present freedoms and past oppression.

A focus on the relationship of democratic structures to past authoritarian eras is a productive approach to the German debates about the exhibition. Since the fall of the Berlin wall and German unification, cultural commonplaces with respect to German democracy have been undergoing shifts and redefinition, meaning that the status and health of German democracy occupy the forefront of many discussions. In recent years, German politicians have felt compelled to remind their constituents of their historical burdens. As Socialist politician Thorsten Meyerer explained in 2001,

> That is the lesson of Auschwitz-Birkenau, of Majdenek and Treblinka, of Sobibor, Belzek and Chelmno: that one is not only responsible for what one does, but rather also for that which one allows to happen.... True, not everyone looked the other way [in the Nazi era]. There was resistance from communists, unions and social democrats, from church groups and individuals such as Dietrich Bonhoeffer. Even in a dictatorship it was possible to decide in favor of humanity over inhumanity. How much simpler must it be for us, living as we do in a free-democratic constitutional state, to stand up and make it clear to Neo-Nazis that society will not accept it when human dignity is breached.[15]

Much like Wetz and Tag, these politicians articulated anxieties about the condition of Germany's democracy, though the conscious, not to mention conscientious, discussion of the health of German democracy speaks more to its relative strength rather than its fragility.

Concerns about the robustness of Germany's democratic traditions were not out of place, however, following German unification in 1990. In 1999 alone, some 10,000 right-wing attacks on Turks, Jews, and historical monuments took place across Germany. In 2000, the number had risen to an estimated 14,000.[16] In these same years, a newly united Federal Republic began to debate whether the time had come for Germany to act more like a "normal" nation in spite of having anything but a "normal" past. The contentiousness of the debates on topics ranging from the wisdom of transferring the federal capital from Bonn to Berlin, the appropriateness of having German military forces serve in Kosovo and Afghanistan, and the building of a Holocaust memorial to the victims of National Socialism demonstrated that German society could not simply turn away from the legacies of the Nazi past.[17] But it has also become clear that other parameters of that engagement are shifting, for instance, in the growing public debates on formerly taboo topics such as German suffering at the hands of the allies in the Second World War, or in Botho Strauss' and Martin Walser's protests of the political, ethical, and aesthetic burdens borne by Germans with respect to the legacies of the Nazi past and the Holocaust.[18]

Because they turn on the valences and cultural functions of what it has meant to Germans to see images of bodies and corpses in public since 1945, the terms of Walser's protest are worth examining. As Walser put it in a 1979 essay, "One glance at an Auschwitz picture suffices and everybody admits at least to himself: we are not finished with this. Regardless of what you do with this, you cannot delegate it."[19] It is worth recalling that one of the strategies employed by the allies after the Second World War to impress upon Germans the magnitude of the crimes committed in the Holocaust was to require them to view piles of corpses in concentration camps such as Buchenwald or photographs of similar carnage.[20] By 1998, the time he gave his (in)famous speech in Frankfurt's St. Paul's Church, Walser had apparently tired of having to confront images of corpses on a regular basis, and he articulated a desire to be able to look away or escape from those images and, by implication, the Nazi past.[21] Such a looking away should no longer be frowned upon, Walser argued, because, for one thing, the treatment of the Nazi past via images of corpses had by 1998 become little more than an empty ritual and, for another, Germans had in fact already learned enough from the past.[22] Though it is eminently possible to take issue with Walser, I would, for now, like to stress two points. First, the emergence of his position makes it clear that the taboo zones defining the relationship of the culture and politics of the Berlin Republic to the Nazi past were shifting, with new fault lines emerging on the question of whether Germans had "learned enough from the past." Second, the public presentation of images of corpses and bodies in Germany is charged in particular ways.

The particularity of the German context can be clarified by taking a look at the reception of Body Worlds during its London showing from March 21, 2002 to February 9, 2003. As when it appeared in Germany, Body Worlds provoked intense controversy in Britain, with the media not hesitating to sensationalize the public presentation of corpses as a "Victorian freak show." As in Germany, opposition came in particular from religious leaders believing that von Hagens showed disrespect toward the dead and from various groups objecting to the commercialization of the human body. Nonetheless, the show touched different nerves in Britain. Much of the controversy in the United Kingdom swirled around specifically British concerns regarding the public display of bodies and the rights of patients. For one thing, Body Worlds spoke to the work of contemporary British artists such as Damien Hirst, Marc Quinn, Rick Gibson, and Anthony Noel Kelly, each of whom linked art and anatomy by aestheticizing human and animal body parts in provocative and sometimes illegal ways.[23] Since early 2001, the British public had also been reeling from the Alder Hey Report, an account of how, in the 1980s and 1990s, a pathologist, Dr. Dick van Velzen, had secretly harvested organs from over 800 dead children in a Liverpool hospital without having obtained proper parental consent. The Alder Hey Report inflamed the public, igniting fears of modern-day body snatching and impropriety among medical practitioners.

Britons readily associated Body Worlds with the Alder Hey scandal, partly because parents of the Alder Hey children attacked the exhibition in the media and partly because von Hagens and his corpses resonated with the images of a body-snatching mad scientist, particularly after von Hagens performed a public autopsy live before paying and television audiences.[24] Insofar as this performance was the first public autopsy in Britain since the Anatomy Act was passed in 1832 (it was also his first anywhere), von Hagens appeared to deliberately violate specifically British taboos and sensibilities.[25] In response, the British press anointed von Hagens "Dr. Frankenstein," likening him to the fictional doctor inspired by the body snatchers of the early nineteenth century.[26] Allegations that von Hagens had illicitly received bodies of prisoners, the homeless, and the mentally ill from an anatomical institution in Novosibirsk fed associations with anatomical horrors of the British past and present.[27]

The issues of legality and democratic values figured differently in the British reception of the show than they had in Germany. The U.K. government responded unenthusiastically, delaying approval to the show while determining whether the displays violated the 1832 Anatomy Act or presented a public health risk, and, in the case of the public autopsy, whether von Hagens had the proper licenses to dissect a human corpse.[28] When the Health Ministry eventually ruled that no existing statute outlawed the show, several MPs threatened legislation against any and all public displays of corpses.[29] Although nothing came of these efforts, von Hagens and his German supporters protested what they regarded as politicians' un-democratic and un-British attitudes, in the process saying as much about their views of democracy as about the British political climate. A defiant von Hagens stated, "I won't be stopped. I am in a democratic country here [in the United Kingdom] and can do anything permitted by law. I am appalled at [their threat of imprisoning me]."[30] German philosopher Franz Joseph Wetz complained, "I'm astonished and irritated that this exhibition has run into such big obstacles in the United Kingdom, where liberalism originated."[31] Officials had, to be sure, failed to embrace Body Worlds, but in the end, Wetz' vision of populist democracy seems to have prevailed: Britons were largely left to decide for themselves what to think about von Hagens' project. In a rash of articles and letters to the editor, supporters lauded the show for democratizing medical knowledge and enlightening the public, while opponents railed against the commercial, sensationalized aspects of the presentation.[32] For all the polarization, few commentators saw the health of British democracy at stake in the show. An exception is "Jenkins."

The situation was different in Germany, where the show was interpreted along the shifting fault lines demarcating the authoritarian past and the democratic present. Stefan Ruzowitzky's film *Anatomy*, the most successful German film of 2000, helps position Body Worlds with respect to Germany's changing cultural and political terrain, especially the legacies of the Nazi past. *Anatomy*, starring *Run Lola Run*'s Franka Potente, capitalizes on and interprets Body Worlds in terms of medical ethics and Germany's Nazi past. Set in the anatomical institute in Heidelberg where von Hagens developed plastination, the film operates in the genre of Gothic horror, making it primarily about the return of the repressed and the uncovering of disavowed history and experience.[33] As such, *Anatomy* depicts the efforts of Potente's character Paula Henning to expose a secret medical society that murders medical students and others with rare diseases in order to produce innovative, unprecedented anatomical specimens. The secret medical society, known as the "Anti-Hippocratic League" for its carrying out research without regard for patient well-being, maintains continuities with experiments carried out by physicians in the Nazi era. In the film, those experiments take the form of victims being injected with a chemical (promidal) that arrests decay and permits dissection and shaping of their living bodies as their tissues harden and die. As Paula learns, the chemical and techniques were developed by her grandfather in order to perform radical anatomical research in the Nazi era, making its legacies an inescapable part of her family and scientific heritage. As Ruzowitzky puts it, "we tried to get across the idea that ... the foundations of this modern university building lie in the past, a somewhat evil past at that. When we scratch at the surface of modern technological medicine we reveal the sins of our forefathers."[34] By emphasizing the ongoing use of promidal to further illicit research, the film suggests that the institute's spectacular displays of human anatomy are the immediate heirs to Nazi practices, with lucrative contracts from pharmaceutical companies and international prestige driving present-day "anti–Hippocratic" practices in anatomy and medicine.

The film's commentary is unmistakably directed at von Hagens' Body Worlds. Prominent scenes take place in a collection of anatomical specimens, which sports skulls, organs suspended in spirits, and freestanding figures in poses taken straight from Body Worlds. When a student asks, moreover, whether the latter are merely models, he is told, "No, [they are] genuine human specimens, plastinated [*plastiniert*]. All the water and fat is sucked out of the tissue. They'll keep forever."[35] As Ruzowitzky notes, a "big coup in terms of props was our hall of anatomical specimens. The specimens are what the film is all about, really. They were inspired, of course, by that Body Worlds exhibition which is such a big hit these days."[36] Though the plastinated specimens are easily exploited to support the film's presentation of horror and gruesome images, it is significant that they result from and continue to provoke violence and crime. Standing before the plastinated remains of her former roommate Gretchen, Paula realizes that the removal of identifying features such as facial skin — a key means by which Body Worlds specimens are universalized — works only to erase evidence of murder. The cutting away and arrangement of tissue and the contorted poses characteristic of plastinates are thus interpreted as markers of violence and torture. Even more crucially, the bodies on display are presented not as freely donated gifts to science, but as the victims of scientists who disregard basic ethical tenets.

In leveling the charge that universalizing his bodies conceals a crime, the film joins a series of critics who have accused von Hagens of shady dealings and body snatching, faulting him in particular for refusing to present documentation to establish that he obtained his bodies legally and ethically.[37] These allegations reached a fever pitch in 2004, when *Der Spiegel* uncovered evidence that von Hagens' facility in China had accepted the bodies of executed Chinese prisoners. Though von Hagens relinquished the bodies and has not (yet) been convicted of any crime, the episode made it harder for him to continue exhibiting in Europe. In contrast to critical interventions such as *Der Spiegel*'s, however, the film does not shy away from linking these concerns about Body Worlds directly to the dubious practices of the Nazi era. Body Worlds serves as the vehicle for suggesting that modern medicine is accompanied by abuses we would prefer to believe ended in 1945. In this view, the exhibition represents not as a triumph of enlightened science, but rather as its horror-inspiring abject.

Anatomy's association of Body Worlds with Nazi medicine garnered criticism because for many, such a topic seems inappropriate for an entertainment film. The scene in which Paula's Professor Grombek explains the afterlife of the National Socialist medical practices occurs in the basement chambers of the Anti-Hippocratic league (predictably enough, the chambers are the foundation of the medical faculty). As Ruzowitzky notes, "the scene is meant to give the intellectual context, to show that it's not all about going at people with knives.... In this scene, [the Professor] talks about the Anti-Hippocratics in the Third Reich. For some reason, people objected to this. They did not think Mengele and the medical experiments of the Third Reich should be mentioned in a "light" movie like this. I found that a bit odd. I'd be more critical of the movie if it sidestepped the issue. How can you make a movie in Germany about medical ethics and not mention the medical experiments of the Nazi era?"[38] The nature of these objections raises two points. First, the film's view of Body Worlds as heir to Nazi science was not the basis of the objection per se, as it might have been if public perceptions of the exhibition saw no possible association of von Hagens' exhibition with Nazi practices. Second, since unification, the legacies of the Holocaust have increasingly been relegated to "serious" art forms and discourses.[39] *Anatomy*'s violation of this expectation points to the constraints, social as well as discursive, that limit German discussions of the Holocaust.

Andreas Nachama's case further clarifies these discursive limits, showing how Body Worlds inscribes discursive taboo zones with respect to the Nazi past. Nachama, the President of the Jewish Community in Berlin, explained in an interview that the Jewish Community had refrained from mounting an official protest of Body Worlds, because its pronouncements are readily associated with Holocaust issues and the weight of Auschwitz.[40] But Nachama expressed his personal reservations about Body Worlds, stating that the exhibition was "possibly the logical consequence of what happened in the twentieth century. Then there was no barrier to [what could be done] to living human beings, human bodies by the millions were burned to ashes or turned into soap by murderers, lampshades were made out of human skin.... These people deserve to be buried because their earthly existence has ended. They should not be exhibited as a commodity."[41] Christian churches have made similar points about the right to proper burial, the protection of human dignity, and the commodification and instrumentalization of the body in the show without provoking ire.[42] The significant difference between their critiques and that of Nachama is the association of the exhibition with the Holocaust. Yet for von Hagens and a few subsequent critics, this association seemed to violate an unspoken rule.

Rather than take the critique in stride, as he has tended to do with respect to the Christian organizations, von Hagens lashed out at Nachama with an open letter in the *Frankfurter Rundschau*. Differentiating his shows from Nazi science on the sole grounds that his subjects donated their bodies willingly—an argument von Hagens can no longer easily make in light of the Chinese scandal—von Hagens concluded that his exhibition represented "the opposite of the horror of mass murder and the piles of corpses in German concentration camps."[43] Pointing out that he is an "historically aware German, who has visited both Dachau and Buchenwald," von Hagens warned Nachama that his critique tended to trivialize the Holocaust and the suffering of its victims, and "damages the reputation of the Jewish Community as a moral institution that is to be taken seriously."[44] Even giving von Hagens the benefit of the doubt, his disproportionately angry defense is based on evidence that only indirectly acquits his show of any relationship to the Holocaust, if it does so at all. By invoking Dachau and Buchenwald in particular, von Hagens appeals to a ritual that remains meaningful only when it does not foreclose the possibility of future critical self-examination, an exercise von Hagens does not always seem capable of. His response suggests that a taboo is violated when the Holocaust is mentioned in the same breath as his show.

In discussions of Body Worlds that avoid mentioning the Holocaust, whether supportive or critical of the exhibition, the German Constitution does the "work" of anchoring the argument in a way that seems not to exceed certain discursive limits. Though the Constitution is just one step removed from the Holocaust, that level of displacement demarcates a taboo zone that can be crossed only in particular circumstances. Reinforcing discursive separations from the Nazi era would seem to be one crucial factor why debate over Body Worlds comes back again and again to the German Constitution and what German legal principles permit. The invocation of the Constitution can of course be rhetorically complex, even within a single argument. Nonetheless, invoking the Constitution forces an opponent to run the risk of violating at least its spirit, a tricky maneuver not least because such a violation can appear to be ambiguous with regard to past totalitarian eras. The reenactment of political breaks with respect to Nazi and GDR pasts underscores the notion that German democracy exists in some registers as a ritual avoidance of the past, even several decades after the founding of the Federal Republic. Seen this way, the ritual "work" done by the Constitution takes the form of an affirmative gesture toward "democratic values" even as it forestalls open scrutiny of the past.

In conclusion, it is worth recalling that traditions can be renewed in a democracy through a critical exercise of reason coupled with an implicit rejection of earlier non-enlightened, non-democratic traditions. While the critical evaluation of traditional behaviors may be an enlightening exercise, decisions to reject the past can also be driven by non-rational impulses. In the case of post-unification Germany, the disentangling of past and present is loaded with symbolism that also needs to satisfy emotional exigencies, for instance the deep desire to differentiate the present from the Nazi past. In this context, Body Worlds offers a setting that permits visitors to dissociate past and present in a customary fashion, which is to say, both socially shared and historically conventional. It is therefore significant that, historically, techniques of exhibition have been extremely well equipped to transmit a sense of culture as rupture with the past, thus emphasizing symbolic claims to uniqueness, innovation, and progress. In his analysis of nineteenth-century World Expositions and smaller popular exhibitions, Tony Bennett contends that they "located their preferred audiences at the very pinnacle of the order of things they constructed. They also installed them at the threshold of greater things to come."[45] Body Worlds' multiple claims to originality, progress, and the breaking of taboos share much with their nineteenth-century brethren. Body Worlds might therefore be serving German democracy not only in a straightforward, educational fashion, but also by calling forth a ritual affirmation of values antithetical to those of the National Socialists and to a lesser extent the GDR. Precisely because Body Worlds operates in a discursive realm whose terms are decisively "beyond" the Holocaust, visitors to the show are prone to enforce a sense of difference with respect to the Nazi and GDR pasts. This occurs both in the sense of rupture provoked by plastination, as well as in the propensity of the exhibition to generate polarized interpretations. Ultimately, the urgent appearance of democracy at this juncture, in a practice that confirms Germans' perceptions of cultural progress, emerges with a certain logic, even as the exhibitionary apparatus that underwrites those perceptions hails from another era.

Notes

1. von Hagens, "Anatomie und Plastination," 28–29.
2. Brock, Interview, 295–6.
3. Wise, *Capital Dilemma*, 23–38, 121–34.
4. Phillips, "New Art for the Reichstag," 27.
5. Ibid., 27. I am grateful to Daniel J. Sherman for pointing out this connection.
6. Assheuer, "Die Olympiade der Leichen," paragraph 11.
7. Tatlow, "Dead Body Shop," 5.

8. Ibid.

9. Roth, "Menschenökonomie oder der Mensch als technisches und künstlerisches Meisterwerk," 39–67.

10. Charlton et al., "Was die Menschen an den 'Körperwelten' bewegt," 330.

11. Institut für Plastination, "kommentare."

12. Li, "Body of Evidence."

13. Assheuer, "Die Olympiade der Leichen. Der Tabubruch erreicht eine neue Qualität: Gunther von Hagens' Körperwelten ziehen in ein Hamburger Erotik-Museum."

14. Wetz and Tag, "Körperwelten," 15.

15. Meyrer, "Rede anlässlich des Holocaust-Gedenktages am 27.01.2001."

16. Thierse, ""Ansprache von Bundestagspräsident Wolfgang Thierse bei der Veranstaltung im Deutschen Bundestag zum Gedenktag für die Opfer des Nationalsozialismus."

17. Huyssen, Werner Jung, and Peter M. McIsaac, "Introduction," 3–8.

18. Strauss, "Anschwellender Bocksgesang," 37–51.

19. Walser, "Auschwitz und kein Ende," 30–31; translated by Hell in "Eyes Wide Shut, or German Post-Holocaust Authorship," 14–15.

20. I am grateful to Evelyn Preuss for bringing this practice to my attention.

21. Hell, 14–15.

22. Walser, "Erfahrungen beim Verfassen einer Sonntagsrede," 37–51.

23. Jeffries, "The Naked and the Dead"; Connor, "Body of Criticism Greets Artist's Human Display"; Campbell-Johnston, "You've Got To Admire His Guts."

24. "Alder Hey Group's Outrage," 5; Jeffries; Tatlow, 5.

25. Richardson, *Death, Dissection, and the Destitute*, xxiii–xvii.

26. Jeffries; Cavendish, "I Will Cut Up My Father, But Even I Couldn't Do It to My Own Wife," 25–6.

27. Harris and Berlin, "Trade in Bodies Is Linked to Corpse Art Show: Behind 'Dr. Frankenstein's' Exhibition."

28. "Alder Hey," 5; Millward, "Health Department Tries to Ban Corpse Exhibition," 11; Jenkins, "A Bit of a Freak Show, But Where's the Harm?" 11.

29. "Alder Hey," 5; Brownell and Binlot, "Science or Sideshow?"; "Controversial Body Show not Breaking Any Law," 19.

30. "Leichenspektakel: Professor besteht auf Obduktion."

31. Connor, "Body of Criticism," 9.

32. Grant, "Value of Human Corpse Exhibition"; Jenkins, 11; Sacker, "Give Death a Chance," 2; Threadgold, "Body Exhibition Is Education, Not Art," 29.

33. Hantke, "Germany's Secret History: Stefan Ruzowitzky's *Anatomie* (*Anatomy*, 2000)."

34. Ruzowitzky, "Director's commentary."

35. *Anatomy*.

36. Ruzowitzky.

37. Assheuer.

38. Ruzowitzky.

39. Hake, *German National Cinema*, 182.

40. Nachama and Bernitt, "'Jetzt kommt der Spaß am toten Körper.' Interview mit Andreas Nachama."

41. Ibid.

42. Broder, "Gunther von Hagens."

43. Berger-Lenz, "Körperwelten. Ausstellung und Einstellungen"; Diehl and Sundermeier, "Hagens' Leichen"; von Hagens, "Open Letter to Andreas Nachama."

44. von Hagens, "Open Letter to Andreas Nachama."

45. Bennett, *The Birth of the Museum: History, Theory, Politics*, 82; See also Macdonald, "Exhibitions of Power and Powers of Exhibition: An Introduction to the Politics of Display," 12.

Bibliography

"Alder Hey Group's Outrage at Corpse Exhibition." *Daily Post*, March 12, 2002.

Anatomy. DVD. Directed by Stefan Ruzowitzky. Culver City, CA: Columbia Tristar, 2000.

Assheuer, Thomas. "Die Olympiade der Leichen. Der Tabubruch erreicht eine neue Qualität: Gunther von Hagens' Körperwelten ziehen in ein Hamburger Erotik-Museum." *Die Zeit*, August 21, 2003. http://www.zeit.de/2003/35/K_9arperwelten?page=1 (accessed February 14, 2008).

Bennett, Tony. *The Birth of the Museum: History, Theory, Politics*. New York: Routledge, 1995.

Berger-Lenz, Monika. "Körperwelten. Austellung und Einstellungen." *Faktuell*, April 4, 2001. http://www.faktuell.de/Hintergrund/Background116.shtml (accessed February 14, 2008).

Brock, Bazon. Interview by Nina Kleinschmidt and Henri Wagner, in Kleinschmidt and Wagner, 294–6.

Broder, Henryk M. "Gunther von Hagens." Henryk M. Broder. April 23, 2001. http://www.henryk-broder.de/html/schm_hagens.html (accessed February 14, 2008).

Brownell, Ginanne, and Ann Binlot. "Science or Sideshow?" *Newsweek (exclusive web edition)*, March 21, 2002. Lexis Nexis Academic, Reed Elsevier Inc. http://www.newsweek.com/id/75035 (accessed February 14, 2008).

Campbell-Johnston, Rachel. "You've Got To Admire His Guts." *The Times*, March 22, 2002.

Cavendish, Lucy. "I Will Cut Up My Father, But Even I Couldn't Do It to My Own Wife." *The Evening Standard*, November 18, 2002.

Charlton, Michael et al. "Was die Menschen an den "Körperwelten" bewegt." In *Schöne neue Körperwelten: Der Streit um die Ausstellung*, edited by Franz Josef Wetz and Brigitte Tag, 328–356. Stuttgart: Klett-Cotta, 2001.

Connor, Steve. "Body of Criticism Greets Artist's Human Display." *The Independent*, March 23, 2002, 9.

"Controversial Body Show not Breaking Any Law." *Belfast News Letter*, March 21, 2002, 19.

Diehl, Verena Sarah, and Jörg Sundermeier. "Hagens' Leichen." *Jungle World*, April 4, 2001. http://www.nadir.org/nadir/periodika/jungle_world/_2001/15/19a.htm (accessed December 31, 2003).

Eshel, Amir. "Vom einsamen Gewissen. Die Walser-Bubis Debatte und der Ort des Nationalsozialismus im Selbstbild der Berliner Republik." *Deutsche Vierteljahrsschrift für Literaturwissenschaft und Geistesgeschichte* 74 (2000): 333–60.

Grant, Gary. "Value of Human Corpse Exhibition." *The Times of London*, March 27, 2002.

Hake, Sabine. *German National Cinema*. New York: Routledge, 2002.

Hantke, Steffen. "Germany's Secret History. Stefan Ruzowitzky's *Anatomie* (*Anatomy*, 2000)." *Kinoeye: New Perspectives on European Film* 1, no. 1 (2001). http://www.kinoeye.org/01/01/hantke01.php (accessed December 31, 2003).

Harris, Paul, and Kate Connolly Berlin. "Trade in Bodies Is Linked to Corpse Art Show: Behind 'Dr. Frankenstein's' Exhibition." *The Observer*, March 17, 2002.

Hell, Julia. "Eyes Wide Shut, or German Post-Holocaust Authorship." *New German Critique* 88 (2003): 9–36.

Huyssen, Andreas, Werner Jung, and Peter M. McIsaac. "Introduction." *New German Critique* 88 (2003): 3–8.

Institut für Plastination, "e.K. Besucherkommentare." Koerperwelten: Die Faszination des Echten. Heidelberg: Institut für Plastination, 2002. http://www.koerperwelten.de/en/kommentare.htm (accessed October 5, 2002).

Jenkins, Simon. "A Bit of a Freak Show, But Where's the Harm?" *The Evening Standard*, November 21, 2002, 11.

Jeffries, Stuart. "The Naked and the Dead." *The Guardian*, March 19, 2002.

Kleinschmidt, Nina, and Henri Wagner. *Endlich unsterblich? Gunther von Hagens — Schöpfer der Körperwelten*. Bergisch Gladbach: Lübbe, 2000.

"Leichenspektakel: Professor besteht auf Obduktion." *Der Spiegel*, November 20, 2002. http://www.spiegel.de/panorama/0,1518,223538,00.html (accessed November 20, 2002).

Li, Ho Ai. "Body of Evidence: Plastic Fantastic." *Life! The Straits Times*, November 13, 2003. Article posted by Body Worlds in 2004 at http://www.bodyworlds.com.sg/en/news-01.htm (accessed June 27, 2004).

Macdonald, Sharon. "Exhibitions of Power and Powers of Exhibition: An Introduction to the Politics of Display." In *The Politics of Display: Museums, Science, Culture*, edited by Sharon Macdonald, 1–24. New York: Routledge, 1998.
Meyrer, Thorsten. "Rede anlässlich des Holocaust-Gedenktages am 27.01.2001." SPD Kreisverband Miltenberg. http://www.spd-miltenberg.de/Rede_Holocaustgedenktag.html (accessed February 14, 2008).
Millward, David. "Health Department Tries to Ban Corpse Exhibition." *The Daily Telegraph*, March 12, 2002, 11.
"MP Condemns 'Ghoulish' Display of Treated Corpses." *Western Mail*, March 19, 2002, 7.
Nachama, Andreas, and Sven Bernitt. "'Jetzt kommt der Spaß am toten Körper.' Interview mit Andreas Nachama," *Die Welt am Sonntag*, March 11, 2001.
Paret, Peter. *The Berlin Secession: Modernism and Its Enemies in Imperial Germany*. Cambridge: Belknap, 1980.
Phillips, Christopher. "New Art for the Reichstag." *Art in America* 87 (1999): 27–28.
Roth, Martin. "Menschenökonomie oder der Mensch als technisches und künstlerisches Meisterwerk." In *Der gläserne Mensch, eine Sensation. Zur Kulturgeschichte eines Ausstellungsobjekts*, edited by Rosmarie Beier and Martin Roth, 39–67. Stuttgart: G. Hatje, 1990.
Ruzowitzky, Stefan. "Director's commentary." *Anatomy*. DVD. Directed by Stefan Ruzowitzky. Culver City, CA: Columbia Tristar, 2000.
Sacker, Max. "Give Death a Chance." *The Independent*, March 28, 2002, 2.
Strauss, Botho. "Anschwellender Bocksgesang." *Der Spiegel*, February 8, 1993.
Tatlow, Didi Kirsten. "Dead Body Shop." *The Sunday Business Post*, August 5, 2001.
Thierse, Wolfgang. "Ansprache von Bundestagspräsident Wolfgang Thierse bei der Veranstaltung im Deutschen Bundestag zum Gedenktag für die Opfer des Nationalsozialismus." Address Before the German Bundestag, January 26, 2001. http://www.bundestag.de/aktuell/presse/2001/pz_010126.html (accessed February 14, 2008).
Threadgold, Steve. "Body Exhibition Is Education, Not Art." *The Daily Telegraph*, March 22, 2002, 29.
von Hagens, Gunther. "Anatomie und Plastination." In *Prof. Gunther von Hagens' Körperwelten. Die Faszination des Echten*, edited by Angelina Whalley, 22–35. Heidelberg: Institut für Plastination, 2000.
_____. "Open Letter to Andreas Nachama." *Frankfurter Rundschau*, March 22, 2001.
Walser, Martin. "Auschwitz und kein Ende." In *Über Deutschland reden*, edited by Martin Walser, 24–31. Frankfurt am Main: Suhrkamp, 1988.
_____. "Erfahrungen beim Verfassen einer Sonntagsrede." In *Friedenspreis des deutschen Buchhandels, 1998*, edited by Martin Walser, 7–28. Frankfurt am Main: Suhrkamp, 1998.
Wetz, Franz Josef, and Brigitte Tag. "'Körperwelten.'" In *Schöne neue Körperwelten. Der Streit um die Ausstellung*, edited by Franz Josef Wetz and Brigitte Tag, 7–20. Stuttgart: Klett-Cotta, 2001.
Whalley, Angelina, ed. *Prof. Gunther von Hagens' Körperwelten. Die Faszination des Echten*. Hiedelberg: Institut für Plastination, 2000.
Wise, Michael. *Capital Dilemma: Germany's Search for a New Architecture of Democracy*. New York: Princeton Architectural Press, 1998.
Zuschlag, Christoph. *"Entartete Kunst": Ausstellungsstrategien im Nazi-Deutschland*. Worms: Wernersche Verlagsgesellschaft, 1995.

The Echo of German Horror Films

Alexandra Ludewig

This essay highlights the confluence of the re-birth of the German horror film genre, the associated controversial re-emergence of thematic references to Nazism, and the role played by Gunther von Hagens' Body Worlds phenomenon in a number of films subscribing to this genre. An examination of this confluence reveals much about human nature — about our collective tendency to do evil by neglecting what is good — by succumbing to authoritarianism in its many forms, and by failing to resist the ever-looming threat of totalitarianism in the mistaken belief that Nazism and its like are safely located in the past. The paper mounts a defense of a theory about the nature of evil — a theory that is exemplified in Stefan Ruzowitzky's *Anatomy* films (released in 2000 and 2003) and that is pertinent to the work of von Hagens. Evil is not exclusively couched in the personae of a von Hagens or horror film character inspired by him, but rather it is a phenomenon of apathy or fear: the potential for wrongdoing lies in all of us. As such it is the human inclination to submit to authoritarianism that can facilitate widespread evil.

The work of Hannah Arendt, particularly her book *Eichmann in Jerusalem* (1963), and Stanley Milgram's findings in *Obedience to Authority* (1974), which resulted from a series of experiments conducted in the early 1960s, and, which, like Arendt's book was inspired by the behavior of Adolf Eichmann, will set the theoretical background. Both Arendt and Milgram point to a human propensity to engage in evil under certain conditions. Although good and evil are abstractions made by individuals or societies at certain times, both Arendt and Milgram identify the existence of pandemic evil — whether a specific phenomenon such as Nazism, or a generalized notion as in the work of Arendt — as a species of totalitarianism that is facilitated by submission to authority. Arendt and Milgram identify these authoritarian figures, most often present in patriarchal authority, as being endemic to totalitarian systems such as Nazi Germany and Stalinist Russia, or in institutional authority as exercised by scientists and academics.[1] Both Arendt and Milgram demonstrate that individuals who exercise their power through charisma and organizational legitimization can go astray, but, more worryingly, lead others astray. It is precisely because their deeds remain unchallenged that they are perceived to be God-like figures who "know what's best for you,"—a confluence that highlights the flaws of human nature.

An examination of Stefan Ruzowitzky's horror films, which are as much a social commentary on von Hagens and his Body Worlds exhibits as they are an analysis of evil in society, helps to illustrate one element of Arendt's and Milgram's theories—that, while under authoritarian regimes, the tendency to commit atrocities against other human beings is commonplace rather than exceptional. The films confirm what Arendt calls the "banality of evil."

While German-speaking filmmakers are regarded as co-founders of the horror genre, having shown their mastery with many silent films considered to be expressionistic classics, the genre in German-language film was actually in decline by the mid–1920s. The situation worsened after 1945, with Nazi history rendering on-screen horror inappropriate—an unnecessary reminder of the "bad German." It has taken over sixty years for German-speaking directors to bring the horror genre back to cinemas; successful examples include *Das Experiment (The Experiment)* by Oliver Hirschbiegel (2001), which was also directly inspired by Milgram's work, *Anatomie (Anatomy)* by Stefan Ruzowitzky (2000) and its sequel *Anatomie 2* (2003), as well as Robert Schwentke's *Tattoo* (2002). All three directors have created films that reconnect with the pre–1938 cinematic tradition and the Nazi period alike.

Stefan Ruzowitzky, however, adds a further level of connection and critique: in the feature film *Anatomie (Anatomy,* 2000) he includes unmistakable references to the performance artist Gunther von Hagens and his "Körperwelten" (Body Worlds) exhibits. Set in Heidelberg, the film exposes a group of medical students and their professional mentors who kill their victims in order to preserve the bodies. Like Dr. von Hagens, who worked at the University of Heidelberg and perfected his techniques there, the character Professor Grombek uses plastination to prepare human specimens and claims that "plastinated organs and body slices are a novel teaching aid."[2] Just as von Hagens has encountered difficulty in explaining the origins of many of the body parts in his exhibitions,[3] Professor Grombek is unable to account for the ethical conduct of his staff and students. In Ruzowitzky's first *Anatomy* film, Grombek's team members abuse their power by performing vivisection on their victims. In the case of *Anatomy 2*, the allure of power and money corrupts the medical scientists.

Other overt connections between Ruzowitzky's films and the accusations leveled at von Hagens include the deliberate public display and celebration of their sophisticated anatomical dissection, as well as the exploitative presentation of human bodies as objects of spectacle and wonder, allegedly in the name of science. In both cases, dissected bodies and body parts are displayed on pedestals, lit eerily, and generally showcased in a way that invites interaction (in the way the models may be opened up like chests of drawers, for instance).

In his film, Ruzowitsky highlights the fact that anatomists have historically operated in murky legal and ethical waters. His *Anatomy* films point to instances when technology has corrupted many a "God in a white coat." In both films, Ruzowitzky examines a fictional "Anti-Hippocratic" society, analyzing its transgressions, which stem from institutional as well as anthropological shortcomings, all the while inviting comparison with the viewers' fascination with naked corpses.

Anatomy (2000) and its sequel *Anatomy 2* (2003) make use of contemporary settings in an attempt to emphasize timeless values, perceptions, and fears. Both films fuse historical references with elements of popular genre and utilize present-day society as a backdrop in order to appeal to a younger viewing public in particular. In so doing, Ruzowitsky has arguably created a new school of films that reveals an unusual perspective on Nazi crimes, not only because of their contemporary setting, but also due to their choice of genre. Ruzowitzky packages Nazi and Fascist themes in a genre which most German and Austrian filmmakers have shied away from as much as they have from films that could be read as portraying Fascism as human. However, Ruzowitzky breaks with these unwritten laws by creating horror movies in the best Hollywood tradition, while suggesting at the end of his films that horror appeals to us all on a very primordial level.

In *Anatomy,* a horror fantasy about the medical profession, Franka Potente plays the protagonist Paula Henning, a promising medical student with a strong background in the field: her grandfather is a celebrated surgeon and her father runs a clinic for people on low incomes. Paula earns an internship at the prestigious Heidelberg University, employer of the brilliant anatomist Professor Grombek, whose professional milieu is arguably representative of that of von Hagens as well as of Hannah Arendt's totalitarian type.[4] From within the elitist circle around Professor Grombek in Heidelberg, Paula discovers the sinister dealings of the "Anti-Hippocratics," a group of doctors who secretly pursue medical research at all costs. Deliberately rejecting the Hippocratic Oath, they perform experiments involving dissection or plastination on unwilling — occasionally still living — victims and, further, commit murders in order to prevent exposure of their illegal acts.

Setting the film in Germany magnifies themes and images that evoke the horrors of the Holocaust, explicitly reminding viewers of the similarities between the cruel operations carried out before their eyes and the medical experiments conducted by the Nazis. The violence meted out by the hypermasculine medico Hein marries murder and sexual pleasure, when he is shown lewdly playing with his victim's breast and exposing her genitals, for example. Additionally, his black lab coat is a visual reminder of the aggression of neo–Nazi groups in Germany during the 1990s as much as of some Nazi uniforms in the 1930s and 1940s. The first victim encountered by Paula is called David, a name generally considered to be Jewish in many European countries, which is thus an overt reference to the similarities between the crimes committed by the "Anti-Hippocratics" and those carried out by the Nazis. The ruthless practices of Nazi doctors, most notably Josef Mengele, are invoked early in the film. Not least due to the chosen city (Heidelberg), the university department involved (Anatomy) and the constellation of an authoritative team leader with a fleet of followers, the associations also extend to von Hagens, who, on the occasion of the opening of his Body Worlds Exhibition in Cologne in 2000, was confronted by protestors bearing a banner reading "Mengele 2000." In von Hagens' case, criticism that his work evoked unwelcome comparison with experiments carried out under Nazism and Stalinism intensified when it became known that one of the bodies in his exhibition had Cyrillic characters tattooed on its lower arm, suggesting that the owner may have been a former prison camp inmate.[5]

In contrast to the controversial reception of von Hagens' work in Germany, which led him to permanently turn his back on his homeland in 2004, Ruzowitsky's critique of anatomical practices in *Anatomy* was extremely well received in Germany and Austria and successfully marketed abroad (the film is doing particularly well in the United States, the United Kingdom, and Australia) by its American distributor, Columbia Pictures. Such was its popularity that a sequel was produced in 2002, an unusual occurrence in German-language cinema, with the German subsidiary company Deutsche Columbia TriStar Filmproduktion endeavoring to cash in further on the success of its first production in Europe. As a result, with regard to visual aesthetics as well as to marketing, *Anatomy* and *Anatomy 2* have a rather American feel about them and derive a lot of their popular appeal from intertextual horror genre conventions encountered in American "teen-horror" and slasher films. These references are especially apparent in the sequel, a much more generic story that does not foreground national iconography, despite its setting in Berlin.

In *Anatomy 2* there are no longer any of the direct references to Mengele or Nazism that underlined the national semantic relevance of the first film, but Ruzowitsky's critique of the authoritarian type and institutional abuse of power continues. Once again, the grim dissection suite of university medical facilities is located in the basement of a labyrinthine building. The medical faculty's underground corridors provide a setting for the realization of fantasies and delusions of grandeur. In this shadowy world, the reigning master is the neurosurgeon Professor Dr. Müller-LaRousse, a prime example of the authoritarian type.[6] He manipulates his colleagues and interns and initially also the film's male protagonist, Jo Hauser, in order to conduct unethical research into muscle transplants and synthetic tissue. The professor's medical advances are designed to broaden his powerbase rather than to help relieve suffering. The prime motivators behind his experiments are his desire for power, influence, and infamy, all of which are guided less by a desire to heal than to improve on nature. His vision is to achieve external control of people by taking over their nervous system: a mad scientist's contemporary Frankensteinian vision based on the latest techniques derived from bio-mechanics and chemistry. Though nominally a member of a humanitarian institution, he is instrumental in causing the medical faculty to sell out to market interests, chiefly insurance companies and pharmaceutical multinationals. This dynamic illustrates how the fundamental ideal of helping people in need is suspended in instances when the patient is unable to produce a valid health care card or the appropriate payment in exchange. Medical care, if provided at all, is based on a business model rather than the requirements of the Hippocratic Oath. Chief physician Müller-LaRousse's neuro-science laboratory thus becomes a site where scientists develop their delusional visions, trying to manipulate nature in the pursuit of creating the prototypical "Übermensch" by experimenting on themselves with drugs and implants. The film's horror is largely derived from the cold detachment evident in the doctors' interactions with the human body; they are reminiscent of Arendt's authoritarian types in *Origins*:[7] the procedural mechanics performed by these medical scientists are contrasted with the human carnage they carelessly cause but which does not weigh upon their conscience.

Müller-LaRousse, as well as Hein and Grombek's other associates, are exposed as figures who prey upon human weakness, on our apparent innate tolerance for the propagation of evil. Even in the face of despicable acts, submission to authority and simple negligence are paramount. If any persuasion beyond the acceptance of authority is deemed necessary, it is applied by invoking the capitalist profit motive as well as an extrapolation of the idea that the death, dismemberment, and violation of the few is permissible so long as it serves the cause of the majority in the pursuit of science and medicine.

The authoritarian objective is further enhanced by the refusal or inability of humankind to learn from history. As such, Ruzowitzky's commentary on history is as poignant as that on the possible future of mankind. The frailty of civil society and its tendency to descend into chaos are integral aspects of any dynamic system. Evidence of the latter had already led the philosophers of the Frankfurt School in the late 1940s and 1950s to ask how it was possible "that in a culture of law, order, and reason, there should have survived the irrational remnants of ancient racial and religious hatreds [leading to] the willingness of great masses of people to tolerate the mass extermination of their fellow citizens."[8] As Arendt suggests in *Eichmann in Jerusalem* (1963), banal carelessness is implicated as a causative factor when unremarkable people become part of a group of aggressors.

Jo Hauser's case in *Anatomy 2* confirms that, within the framework of a totalitarian system, apathy encourages the abuse of power. In both *Anatomy* films, surgeons dissecting bodies or performing procedures on the living chat mindlessly about the mundane. Ruzowitzky thus juxtaposes the horror of the visual imagery, which is enhanced by close-ups of body and scalpel, with the quotidian nature of the dialogue. Nevertheless, many of the doctors remain likable, corresponding with Arendt's rejection of the idea that the majority of SS men were psychopathic sadists.[9] To demonize them and to portray them as natural monsters is precisely to miss the most disturbing fact about the average Nazi perpetrator: namely, that in an overwhelming number of cases they were dutiful, law-abiding citizens who, in their roles as civil servants, carried out orders that they did not question or dare to disobey.

This very point was proved by Stanley Milgram, who conducted a series of experiments on obedience to authority at Yale University in 1961 and 1962. Historically, Milgram's experiments followed the capture of Adolf Eichmann in 1960, coinciding with Eichmann's trial in Jerusalem in 1961, a time when the term "Holocaust" was gaining commonplace usage. Milgram devised the scenario of his experiments in such a way as to answer the question as to whether it was possible that people like Eichmann (or, indeed, American soldiers in Vietnam who executed women and children in the 1960s and 1970s) were just following orders.[10] Milgram found that "almost two-thirds" of his subjects—ordinary local residents—were willing to administer apparently harmful electric shocks to their students on the command to do so by a scientific authority.[11] Both Arendt and Milgram subsequently proved the tendency of "normal" humans to fall victim to totalitarianism. In *Anatomy 2*, Ruzowitzky initially shows Jo Hauser, the film's protagonist, to be a well-integrated, law-abiding member of German society, before outlining how easy it is for him (as for others) to transgress. Just like his fellow colleagues, Jo is not a monster who emerges from outside the normal spectrum of humanity; rather, his shortcomings and mistakes are dramatized as natural behavior. Thus Ruzowitzky stresses that, in an environment in which an authoritarian power structure foments it, pandemic evil is not extraordinary or anomalous, but rather banal and commonplace, thereby commenting less on Nazism and Fascism in particular than on the belief that power corrupts.

Certain elements of a malevolent capitalism —which points inexorably toward the abandonment of care —flourish under such an environment. Again, the reference to von Hagens' exhibition as big business invites the observation that the early idealistic research component of his work evidently takes second place to the financial benefits that can be reaped from body parts as body art. As such, pure science without a social focus is depicted as dystopian in both *Anatomy* films, while implying that von Hagens' touring exhibitions are likewise far removed from Hippocratic ideals. The Body Worlds exhibits emphasize spectacle over science. Von Hagens consistently defends his enterprise as a form of "democratizing anatomy."[12] However, his belief "that it is not necessary to have consent for the use of body parts in every case,"[13] as well as the great personal wealth generated by his exhibitions and his sale of anatomical specimens to institutions across the world, have cast doubts on von Hagens' motives, have prompted some to feel "that the artist is exploiting the dead, stripping them of their dignity posthumously."[14]

Ironically, the accusation of selling out can also be leveled at Ruzowitzky. Despite their allusions to the Nazi regime and Gunther von Hagens, Ruzowitzky's horror movies utilize a generic American formula. Even *Anatomy*'s visual aesthetics are for the most part inspired by stylistic elements familiar from MTV, including fast-paced cuts, swirling camera movements, and a prominent musical score. Ruzowitzky thereby highlights the undeniable Americanization of Austrian and German culture in general and also responds to commercial expectations on behalf of his production company, which clearly had a secondary overseas market in mind. However, Ruzowitzky's films must also be understood as part of a film history in which tentative steps are once again being taken into the realm of horror. His two *Anatomy* films have been welcomed as a successful movement towards this genre, a genre which was to some degree off-limits while under the shadow of post–Nazi self-censorship.

Ruzowitzky's fusion of the elements of the contemporary horror genre with twentieth-century European history is comparable to Robert Schwentke's work in *Tattoo* (2002). In all three films, memories of Nazism and Fascism are evoked by way of body art collections and the integration of sexual arousal, violence, and death, thus simultaneously imbuing the concept of horror with aesthetic, political, and moral values. In *Anatomy*, this connection is accentuated by the group of medical doctors employing Fascist-inspired practices, while the perpetrators in *Tattoo*, not unlike von Hagens, are body art dealers. Biological selection and the relentless pursuit of a science that can aid the creation or survival of the Übermensch (super human), as well as the reference to the "body beautiful," indicate that the Third Reich has left a potent legacy. The actions of body art dealers and the Anti-Hippocratic doctors alike, as well as showmen like Gunther von Hagens, provide a basis for this critique of contemporary society, in which Schwentke and Ruzowitzky see similar ideologies and practices still in existence — against all expectations and despite official denials. A detailed analysis of *Anatomie* illustrates this point.

Even after the exposure of the Anti-Hippocratics' conduct in the present day, the film's heroine Paula finds distinguished professors such as Grombek applauding the secret society's objectives in — as he says in the movie — not restricting "itself to narrow-minded traditions."[15] If this were not the case, it is suggested that Heidelberg University's leading role in the medical field, potential Nobel prizes, government funds for research, commissions from the pharmaceutical industry, and finances to educate "the elite" would be in doubt. Although Grombek admits that during the Nazi era people like Josef Mengele "went awry," he claims that their scientific achievements, specifically in the field of anatomy, have not been given proper recognition: "He admits that they 'play God' while experimenting on terminal patients, but states that the patients do not usually even notice what is going on."[16] As if this claim alone justifies their dealings, he continues to support the actions of professional societies that dare to compete in the global arena, describing them as "unremarkable" in this modern age of cloning and genetic modifications.[17] With this statement by Grombek, Ruzowitzky invites comparisons of the achievements of Nazi-era scientific research with those of present day science.

The conduct of some researchers (such as those looking into artificially enhancing sporting or sexual performance or physical appearance, as frequently highlighted by the media), as well as discussions about euthanasia and Dr. Gunther von Hagens' Body Worlds exhibitions, are but a few representative contemporary examples that make the chilling action in *Anatomy* less fantastic than real. Dr. von Hagens, when employed as a lecturer at Heidelberg University's Institute for Anatomy and Pathology from 1976 until 1994, invented "plastination" in 1978. Since then this technique has become widely accepted in medical circles and is carried out in many institutions throughout the world. Von Hagens attributes this widespread acceptance to "the hard-wearingness, durability, and lifelike state of plastinated specimens, as well as their high instructional value."[18]

Von Hagens—who initially marketed his specimens for public spectacle under a professorial title since revoked by Heidelberg University[19]—claims that his technique of plastination can enlighten the medical profession about the ways in which diseases develop and how lifestyle choices affect health and well-being. Since the late 1990s, von Hagens has made his bodies and body parts available to the public in increasingly daring exhibitions that have attracted audiences totaling more than twenty-five million worldwide, making Body Worlds the most successful traveling exhibition ever created.[20]

Body Worlds highlights the thrill of seeing a partly dissected, naked, sliced or otherwise manipulated corpse — thus playing on people's fascination with death, nudity, and the shock of recognition — and heightens these effects by the use of historically charged locations for the exhibitions, lighting effects, and night-time openings. Von Hagens has managed to stir heated public debate with his provocative presentations, which have included the invitation of strippers to the vernissage of Body Worlds and the exhibition of his works in Hamburg's Erotic Art Museum, elements overtly fusing sex and death. Further, in 2003, riding a float adorned with skeletons, he participated in the annual Berlin Rave Party Love Parade. In general, von Hagens has sought many opportunities to enlist media support for his business ventures.[21] Styling himself visually in the tradition of Medieval and Renaissance anatomists by wearing a hat, von Hagens likes to portray himself as pushing scientific and ethical boundaries like one of his better known predecessors, Andreas Vesalius (1514–1564).

While corpses for dissection were hard to come by even during Rembrandt's time (1606–1669), with authorities only releasing a very limited number of bodies— those of deceased prisoners and criminals— von Hagens has pushed boundaries and broken taboos; the intertwining of sex appeal and death alluded to above is just one example of such. His actions have plunged him into controversy, particularly in his native Germany, where his exhibitions have alarmed audiences more than anywhere else due to the historical burden of the Nazi past. In contrast, many Asian and (since 2004) American critics have heralded his exhibitions as a prime example of the confluence of research and public support.[22]

While Dr. von Hagens has quite intentionally provoked many discussions in Austria and Germany, he has always proudly claimed that each of his plastinated whole-body specimens has been donated voluntarily. Indeed, in recent years many people have explicitly donated their bodies to him in their wills. Nevertheless, there has been considerable controversy surrounding some of the corpses used for the body part exhibits, which feature plastinated hearts and other internal organs, as some of those organs were bought from prisons or mental institutions in countries such as China and Russia.[23] Although strictly speaking this is not illegal, many feel that it is unethical to use people's bodies for artistic purposes without their permission.

The parallels to Ruzowitzky's film are unmistakable. Dr. von Hagens, like Professor Grombek, claims that

> plastinated organs and body slices are an innovative teaching aid for cross-sectional anatomy which is gradually gaining importance and can be easily correlated with radiological imaging. Series of transparent body slices are helpful for manifold scientific research activities. In addition, they are a suitable diagnostic means in pathology, as they allow rapid macroscopic and diagnostic screening of entire organs or operation preparations; they still allow for selective analyses of pathological tissue regions with conventional microscopic methods.[24]

Under von Hagens' purview, human subjects are transformed into specimens and objectified, while the ideals expressed in the Hippocratic Oath are devalued further and further, much as Ruzowitzky's films suggest. Advances in science and the cut-throat competition for the Nobel Prize have become more important than service to humanity; indeed, the scientists' inhumanity is startling and reminiscent of scientific activities carried out under Nazism. Thus, *Anatomy* "explores the fictional premise that underneath the democratic surface of contemporary Germany, and within its technocratic elites, traces of the country's Fascist past still linger."[25]

The film not only highlights a malaise that also applies to other Western societies, it also points out why many Germans seem to be held captive by the spell of the past — a past with which people have not been able to come to terms. An examination of Paula's family helps to shed light on this phenomenon.

Paula is shattered when Professor Grombek reveals that her own grandfather was once the head of the Secret Society, as well as the inventor of the drug "Promidal" that is used in the plastination process. Now an ailing, though still strong-willed man, Paula's grandfather has been her mentor all her life. Even from the confines of his bed in the very hospital in which he used to practice and teach, he has been able to influence Paula's career. Symbolically, and in parallel to her discovery of Grombek's malevolence, Paula's grandfather dies and ceases to be her idol, although he also passes away before she has an opportunity to confront him. With this new insight into her family, Paula is forced to renegotiate her relationship with her father. Thus far, she has disassociated herself from him as she has been unable to understand his life and career choices. He seems to have rejected pure scientific research, fame, money, and ambition — as Paula comes to understand — as a way of protesting against tendencies in the medical world that he perceives to be inhumane. Many years ago, he made the decision to go back to the fundamentals of the Hippocratic Oath (by caring for the needy, the most vulnerable, and least advantaged) as a deliberate and provocative departure from his own father's practices and principles. While his motivations may have been noble, Paula's father has failed to communicate them to Paula. This behavior is depicted as paradigmatic for members of the so called "68-generation," many of whom rejected the world of their fathers outright, without engaging in a personal or differentiated way with the past and their own personal involvement. The film thus suggests that "[t]he grandchildren's generation is susceptible to the proto- or crypto-fascist stance of the Anti-Hippocratics because the previous generation has failed to break this silence."[26] However, the film's critique of German society goes further by implying that Fascist ideology has survived into the present *because* of the failure of the 68-generation.

Anatomy suggests that even its heroine is susceptible to a similar kind of benign ethical failing. When Paula exposes Hein as the culprit, she initially falls victim to the same black and white logic that only sees good or bad; she is unaware that the matter is infinitely more complex. This is evident in the concluding scene of *Anatomy*, which features two of Paula's fellow students talking about their immediate plans following the end of the summer program in Heidelberg. One mentions rumors that a branch of the secret society is still operating in Berlin, while the female student reveals that she has decided to take over her father's practice. She explains that government supervision is not quite as strict in the case of private practices such as her father's, thereby providing some freedom for scientific experimentation. When their conversation is interrupted by the morgue attendant asking how things are going, they respond tellingly that everything is the same as always; it seems that the technocratic elites controlling institutions— however large or small — will continue to provide a safe haven for fascist ideology.

Thus the Austrian-born Ruzowitzky, in a similar manner to Oliver Hirschbiegel in *The Experiment,* exposes institutional as well as anthropological shortcomings that aid the continuation of or digression into fascist practices in Western society. The *Anatomy* films, just like *The Experiment,* present a challenge to conventional ways of reading films about Nazi Fascism because they situate their protagonists within educational facilities. The directors seem to revel in presenting the implosion of contemporary civil society and the regression of members of an educated middle-class to barbarians in a matter of days. As with Hannah Arendt's interpretation of Nazism, the films emphasize the banality of evil, suggesting the degree to which the ordinary person is susceptible to becoming part of the group of aggressors. *The Experiment* and *Anatomy 1* and *2* explore that which "within the individual organism responds to certain stimuli in our culture with attitudes and acts of destructive aggression."[27] In addressing this issue, the films by Hirschbiegel and Ruzowitzky prepare the ground for a broader understanding of evil.

This shift in the representation of evil in Ruzowitzky's *Anatomy* films highlights the successful workings of the confluence of a new depiction of evil and Nazism. In concert with other German films, such as Robert Schwentke's *Tattoo*, Ruzowitzky's foray into the horror genre can be contextualized more broadly within a trend that heralds a change in the understanding of evil, especially with regard to Austria's history and the sustained trend among many Austrians to see themselves as victims of World War II. This highlights the complex nature of humankind, memory, and history in our understanding of good and evil, and sheds light on the representation of evil in contemporary German-speaking culture. Indeed, despite the cinematic shortcomings of the largely generic *Anatomy* films, the novelty value in Ruzowitzky's case is that he has shown a way of moving beyond national modes of explanation in his exploration of evil.

In his re-discovery of the horror genre for German-language film, he has stressed that evil is a timeless, primordial, anthropological tendency that is more or less prevalent depending on a complex mix of conditions. He has thus broadened our perspective and challenged our values and taboos in commonly held representations of evil and of Nazism alike. Ruzowitzky provides his social commentary not in the aesthetics of realism, but instead in the form of a parable. This foregrounds the pervasiveness of evil, the fact that each and every one of us is capable of despicable acts of cruelty. Both *Anatomy* films can thus be read as a reminder of "how a present contemplating a future generally invents the past it deserves."[28] The hyper-graphics and hyper-reality of contemporary cinema in this instance explicitly present the horrors that usually remain off-screen events in films about the Holocaust. While many critics of Hollywood's attempts at producing Holocaust movies have rejected "visual images that saturate and ultimately wipe away the reality of the horror,"[29] Hirschbiegel's, Schwentke's, and Ruzowitzky's images are very graphic. However, their contemporary (if not futuristic) rather than historical settings ensure that it is not the Holocaust that is solved, explained, or commercialized, but its remaining traces in any modern society. Their ultimate message, therefore, is that the Nazi past cannot be dealt with as a historical issue, but that it is part of every contemporary society.

The films deliberately focus on evil and carelessness in the world as constants, thus reminding the audience that crimes associated with the Nazis are in reality crimes associated with humankind. They cannot be relegated to a past period, but must instead be seen as unremitting in the human condition. However cruel the protagonists or antagonists are depicted in the films, they can neither communicate the unspeakable horrors of the past, nor silence the unease about the potential for humanity to do it all over again. Ruzowitzky's contemporary horror films thus consciously accentuate evil through the use of graphic imagery, while providing an explanation of the Holocaust that is removed from the cultural and historical context. By employing ancient motives and contemporary settings, the evil acts are rendered timeless, and the viewer is thus given another point of reference from which to understand the events as anthropological constants:

> Hitler is and remains an anthropological possibility that will occur again and again. That is, so to speak, his legacy. However, this has scarcely been noted up to now. For 200 years after the Enlightenment we had an image of humankind that was based on the assumption: Man is good, Evil can be eliminated through education and improvement of social conditions. Hitler has destroyed this positive image of humankind. We know, or ought to know, that Evil exists ... According to Freud: Under the thin veneer of civilization there is a pack of murderers in every person. An insight which we have to gain once again when we look at the 20th century.[30]

Both *Anatomy* films, as well as *The Experiment* and *Tattoo* can therefore be interpreted not only as allegories of Nazism, but also as commentary on mankind's natural inclination toward totalitarianism. While making viewers wary of humans in general, the films also raise concerns about society's trust in institutions that promote law and order. It is generally believed that anarchy is prevented by the rules and values promoted and upheld by the pillars of society: the church, the family, schools, and universities. However, such horror films seem to imply that it is these very institutions which can aid abuse, by serving as a cover for sinister operations. Nevertheless, these films also show the viewer a way out, by presenting role models who display the courage to stand up for what they believe in. The ultimate truth presented by the films is that people are always responsible for their actions, no matter what situation or historical moment they may find themselves in.

Notes

1. Arendt, *The Origins of Totalitarianism*, passim; Milgram, *Obedience to Authority*, passim.
2. Institute for Plastination, "Preservation by Plastination."
3. "The Anatomists."
4. Arendt, *The Origins of Totalitarianism*, 23.
5. *Der mit den Leichen*.
6. Horkheimer and Flowerman, "Foreword," ix.
7. Arendt, *Origins*, passim.
8. Ibid., 5.

9. Arendt, *Eichmann in Jerusalem*, 286–7.
10. Milgram, *Obedience to Authority*, 5–6.
11. Ibid., 5.
12. von Hagens cited in Bloom.
13. "The Anatomists."
14. Ibid.
15. *Anatomy.*
16. Dolan, *Anatomy aka Anatomie.*
17. *Anatomy.*
18. Institute for Plastination, "Preservation by Plastination."
19. *Der mit den Leichen.*
20. Institute for Plastination, "Prelude."
21. Röbel and Wassermann, "Händler des Todes." 42.
22. Burns, "Gunther von Hagens' Body Worlds," passim.
23. *Der mit den Leichen.*
24. Institute for Plastination, "Preservation by Plastination."
25. Hantke, "Germany's Secret History."
26. Hantke, "Horror Film and the Historical Uncanny," 128.
27. Horkheimer and Flowerman, v.
28. Bosworth, "Film Memories of Fascism," 117.
29. Leventhal, "Romancing the Holocaust."
30. Joachim Fest quoted in Amend, "Was für ein Land?" 57.

Bibliography

Amend, Christoph. "Was für ein Land?" [What kind of country?]. *Die Zeit*, July 10, 2004, 57–58.

"The Anatomists: Anatomy, Ethics, and the Law." PA News. Channel 4, Belfast. http://www.channel4.com/science/microsites/A/anatomists/ethics1.html (accessed January 18 2008

Anatomy. DVD. Directed by Stefan Ruzowitzky. Culver City, CA: Columbia Tristar, 2000.

Arendt, Hannah. *The Origins of Totalitarianism.* New York: Harcourt, 1951.

_____. *Eichmann in Jerusalem. A Report on the Banality of Evil.* New York: The Viking Press, 1963.

Bosworth, Richard J.B.: "Film Memories of Fascism." In *Italian Fascism. History, Memory and Representation*, edited by R.J.B. Bosworth and Patrizia Dogliani, 102–123. London: Macmillan, 1999.

Burns, Lawrence. "Gunther von Hagens' Body Worlds. Selling Beautiful Education." *The American Journal of Bioethics* 7, no. 4 (2007): 12–23.

Der mit den Leichen tanzt — Protokoll eines todsicheren Geschäfts. Die Körperwelten des Gunther von Hagens. Television Documentary. Directed by Christian Schulz and Torsten Peuker. ARD/mdr, arte 2003.

Dolan, Sean Patrick. "*Anatomy* aka *Anatomie.*" Review of *Anatomie* (Sony Pictures). *Cinema Nocturna.* http://www.cinema-nocturna.com/d_anatomy_review.htm (accessed June 12, 2007).

The Experiment. DVD. Directed by Oliver Hirschbiegel. Culver City, CA: Columbia TriStar, 2001.

Fest, Joachim, quoted in Amend, Christoph. "Was für ein Land?" [What kind of country?]. *Die Zeit*, July 10, 2004, 57–88.

Hantke, Steffen. "Germany's Secret History. Stefan Ruzowitzky's *Anatomie* (*Anatomy*, 2000)." *Kinoeye* 1, no. 1 (2001). http://www.kinoeye.org/01/01/hantke01.php (accessed August 14, 2004).

_____. "Horror Film and the Historical Uncanny: The New Germany in Stefan Ruzowitzky's *Anatomie.*" *College Literature* 31, no. 2 (2004): 117–142.

Horkheimer, Max, and Samuel H. Flowerman. "Foreword." In *The Authoritarian Personality*, edited by Theodor W. Adorno et al., i–xxxiii. New York: Harper & Row, 1950.

Institute for Plastination. "Preservation by Plastination." Gunther von Hagens' Body Worlds: The Original Exhibition of Real Human Bodies. Heidelberg: 2006. http://www.bodyworlds.com/en/plastination/idea_plastination.html

_____. "Prelude." Gunther von Hagens' Body Worlds: The Original Exhibition of Real Human Bodies. Heidelberg: 2006. http://www.bodyworlds.com/en/prelude.html

Leventhal, Robert S. "Romancing the Holocaust, or Hollywood and Horror: Steven Spielberg's Schindler's List." Robert Leventhal, University of Virginia. http://www3.iath.virginia.edu/holocaust/schinlist.html

Milgram, Stanley. *Obedience to Authority: An Experimental View.* New York: Harper & Row, 1974.

Röbel, Sven, and Andreas Wassermann. "Händler des Todes. Gunther von Hagens' Ausstellung 'Körperwelten' fasziniert Millionen von Menschen. Doch die vermeintlich künstlerische Präsentation der Toten entpuppt sich als skrupelloses Geschäft: Viele Leichen werden billig gekauft, darunter sind offensichtlich auch die hingerichteter Chinesen." *Der Spiegel*, January 19, 2004, 36–50.

Tatoo. VHS. Directed by Robert Schwentke. München: Universum Film, 2002.

Touching the Corpse: The Unmaking of Memory in the Body Museum*

Uli Linke

The museum is a memory site, a location of commemorative record and practice, where society anchors the past.[1] Unlike other institutions of preservation, the museum collects and displays the material artifacts of a people's cultural heritage: it exhibits history through objects and stages tangible encounters with the past by recourse to the sensually concrete. In the late twentieth century, this materiality or "thingness" of museum exhibits has taken on special significance.[2] For under the impact of global capitalism, in an era marked by the transient, the fugitive, and the contingent, the proximity of objects in the museum tends to promote a heightened sense of anchorage and truth.[3] Moreover, the logic of time and the meaning of memory have been radically transposed. The production of historical consciousness has become increasingly entangled with the commodity form. Encoded by temporal longings, the museum furnishes the common stock for a global memory market. As Huyssen suggests, by transforming historical pasts into cargo-type products, the museum participates in the "marketing of memory" on a transnational scale; in operating as a "cultural memory industry,"[4] it bears the deep imprint of those globalizing forces whereby people's ethno-national remembrances (among others) are commodified, circulated, and consumed. The modern museum is in this sense more than a mere repository of historical artifacts. It is, as de Certeau asserts, a site for the "colonization of time by a discourse of power."[5] Museums not only manufacture new historical archives of peoples, places, and identities, but also participate in a global exchange of memory.

Anthropologists have clearly identified this "traffic in memory" as a key issue during periods of transnational crisis and restructuring.[6] But under globalization,

**This essay originally appeared in* Anthropology Today *21, no. 5 (October 2005): 13–19, and is reprinted here courtesy of Blackwell Publishing.*

with its destabilizing tendencies, we also encounter a new kind of memory market: historical memory as consumer product is increasingly centered on violence and on the body.[7] Global media rapidly appropriate and circulate images of people's suffering: the memories of victimhood are commodified, the remembrance of pain is commercialized.[8] At the same time, the public interest in and market value of bodies has moved the body to the forefront of many museum exhibits. This new prominence of the body museum attests to the formation of a "physiomanic era," as Robert Schenda[9] has phrased it, which emerged "on the threshold of epochal transformations."[10] Under the impact of deterritorialization, simulation, and cyberspace, the very proximity of bodies in the museum satisfies an intense desire for realism and authenticity among a consumer public haunted by the contemporary struggles with memory, history, and temporality.

With a focus on this corporal turn in the museum in mind, here I examine the dramatic staging of Body Worlds (*Körperwelten*), a German exhibition on human anatomy sponsored by the Museum for Technology and Labor in Mannheim in 1997.[11] The installation was launched in 1995 in Tokyo and Osaka, under the auspices of the Japanese Society for Anatomy, where it was endorsed as a scientific exhibition on the art of preserving human remains. Following its German debut in Mannheim, the exhibition has toured through Germany and traveled to a number of major cities internationally, including Vienna, Berlin, London, Seoul, Singapore, and Taipei. Most recently, it was shown in Los Angeles and Chicago. According to media reports from Germany, the exhibition's "dead body specimens" have become "a peak attraction for the general public."[12] In this article I examine this contemporary fascination with corpses. In contrast to previous studies,[13] I focus here on the issue of exhibiting human corpses in Germany, a country in which public culture, despite its affinities with a global modernity, is also framed by issues of nation-state, genocide, and history.

Corpses in the Museum

The objects on display in this exhibition, unlike most specimens in a conventional science museum, are real human corpses: cadavers. The bodies are arranged in various stages of dissection, though without showing any signs of decomposition. The corporeal remains have undergone a preservation procedure known as plastination: the plasticized corpses "do not rot or smell, and they maintain the structure, color and texture of the original tissue and organs down to the microscopic level."[14] Rescued from decay with an "unprecedented measure of realism," the dead are forever preserved in bodily forms that bear the mark of deep anatomical intrusions.[15] The bodies have been flayed, cut open, dismembered, and sectioned to expose the interior world of human physiology.

In this late modern era, dominated by the perpetual simulation of the sensual, the materiality of dead bodies has taken on new meanings. For under the impact of global capitalism, the non-virtual presence of corpses in the museum seems to accommodate a yearning for the thingness of things— the permanent, the tangible, the concrete. This sensory access to corpses might thus nourish a late modern longing for an authentic reality, a perceptual realism without simulation or copies. From this perspective, it is easy to understand why the Body Worlds exhibition "implores the cult of the 'authentic' ... with a perfidious turn to the genuine and real."[16] As curator Gunther von Hagens asserts, "The realism of the specimens contributes enormously to the fascination and power of the exhibit. Particularly in today's media-dominated world ... the exhibition satiates the tremendous desire for unadulterated realism."[17] German exhibition reviews likewise emphasize that the displays are created from "genuine human specimens,"[18] "actual corpses"[19] whose "naked reality"[20] and "authenticity is fascinating,"[21] for the exhibits are "not artificial anatomy models, but real dead people,"[22] "not instructional models, but genuine corpses, presented to the museum visitors openly, in public, and in closest proximity, without the protective glass of the showcase."[23] But the titillating charge that emanates from dead bodies, this fascinating allure of the authentic, needs to be understood in terms of a more general social dynamic. For, as Gottfried Korff points out, "the gaze," or seeing eye, "is also always an expression of a societal self-understanding."[24] Today the quest for anchorage and authenticity is intimately connected to the contemporary turn to the body—a reading which is in some way confirmed by the exhibition's visitor statistics. Altogether sixteen million people have viewed the exhibition worldwide, including eight million in Europe, with six million of these in Germany. Journalists have hailed a "German record for a museum exhibit"[25] with "up to twenty thousand people per day." [26*] The body museum deploys the semiotics of the corpse to appeal to a global consumer public, while at the same time presenting, explaining, and representing the dead in a culturally specific manner.

The "Living" Corpse

The public display of the corpse, which inspires both morbid curiosity and scientific interest, is part of a more general history of exhibitions of the human body, displays in which the themes of death, violence, and medicine are closely intertwined. As Barbara Kirshenblatt-Gimblett notes:

> This history includes the exhibition of dead bodies ..., the public dissection of cadavers in anatomy lessons, the vivisection of torture victims using such anatomical

**These statistics were current at the essay's first printing (October 2005). As of early 2008, total attendance worldwide is estimated to be greater than twenty-five million.*

> techniques as flaying, public executions by guillotine or gibbet, heads of criminals impaled on stakes, public extractions of teeth, and displays of body parts and fetuses in anatomical and other museums, whether in flesh, in wax, or in plaster cast.[27]

Earlier ethnographic or anatomical displays, built around articulated skeletons, taxidermy, wax models, and live specimens, forged a "conceptual link between anatomy and death in the 'museums of mortality.'"[28]* In these museums, human suffering and the horror of death would become interrelated motifs. Body Worlds, however, presents something quite different: it explicitly distances itself from this tradition through its emphasis on unemotional and dispassionate anatomical displays. Its galleries of plastinated corpses are not intended to inspire fear or revulsion by dramatizing death. On the contrary, this collection of dead bodies, which comprises more than two hundred specimens, creates the illusion of life after death. According to Gunther von Hagens, the preservation of the corpse with silicon and plastic makes possible "a new form of posthumous existence,"[29] namely, "the 'resurrection' of the enfleshed body."[30] What are we to make of such statements? And what sorts of images do they invoke?

In this exhibition, human anatomy and physiology are staged through a dramatic tableau of the normal, ordinary, and everyday body. The exhibitors strive for "an aesthetic that strips the dead of their estranging rigidity and instead emphasizes their affinity with the living."[31] Many of these life-sized figures are posed to simulate familiar activities: running, standing, walking, and sitting: "[We see] a basketball player caught mid-dribble with filleted muscles trailing behind ... a human rider mounted nobly on a plastinated horse ... a seated chess player concentrating intently on his game.... A pregnant woman reclining.[32] *The Runner* is frozen in the loping gait of a marathoner."[33]

The bodies on display are staged to perform activities from the mundane world of sports, play, and recreation. In this menagerie of corpses, human remains are presented as "ballerina," "runner," "horse rider," "chess player," "swimmer," and "cyclist." Posed in typified representations with a "proximate realism" to "give the impression that they are forever alive,"[34] these dissected bodies are portrayed as skillfully crafted installations—"the 'living corpses.'"[35] The exhibitors describe their galleries of dead bodies as works of "living anatomy"—not "stiff corpses lying prostrate on the dissecting table," but "entire bodies positioned upright ... standing in life-like poses" that replicate "a living being."[36]

But these presumed works "of authenticity, of living realism in death"[37] are in fact synthetically produced artifacts, mimetic objects. They are designed

**A comparative analysis of the many German anthropological exhibitions containing living and embalmed bodies from the late nineteenth and early twentieth centuries exceeds the scope of this essay. For exemplary references, see Thode-Arora:* Für fünfzig Pfennig um die Welt: Die Hagenbeckschen Völkerschauen, *and Matyssek and Virchow:* Das Pathologische Museum.

to simulate the anatomy of live human beings. And this presentation of vital anatomy is the end product of a series of manipulations, for the bodies are reconstructed, fabricated, aestheticized, and mimetically staged. Nonetheless the exhibitors emphatically insist on the realism of their displays. By circumventing the use of conventional media such as stone, plaster, or wax, they claim to have fashioned human figures directly from organic substances—tissue, muscle, and bone. As such, the exhibited objects are "authentic anatomical specimens."[38] But in the process of manufacture, human corpses are clearly instrumentalized as mere raw material. For the dead flesh is treated, transfigured, and sculpted: "the corpses are worked on without restraint and set in poses."[39] It is an assembly of "human-flesh sculptures."[40]

Sexing the Corpse

The installations embody an ideology of non-alienated flesh, a mythopoeia of corporal authenticity which inevitably seeks to preserve the figures' sexual characteristics. Body Worlds, however, not only articulates the naked sexuality of the corpse: it renders sex iconographically hyper-visible.

The exhibited corpses, with a few exceptions, are male. Every male figure, regardless of its anatomical disarray, has been equipped with an intact penis, emphasizing its sex—its manhood. Even the "deep" body specimens which have been pared down to their muscular or skeletal system show an accentuated display of unharmed male genitalia: enlarged, engorged, and oversized. The exhibits of the female corpse likewise include sexualized poses: "The pregnant figure shields the genital area with her arm, while at the same time framing the foetus in her womb."[41] This woman's curved breasts are prominently accentuated: although her body has been stripped of all its skin, her nipples are placed erect on prosthetic tissue-plates. Indeed, following the exhibition in Mannheim in 1997, this figure was replaced by another female specimen: although displayed in an identical pose, her breasts are conspicuously enlarged. The exhibition of corpses or anatomical "object-bodies," as Horkheimer and Adorno phrased it, requires sexualization, a libidinal charge, in order "to be at once desired as that which is forbidden."[42] The display of corpses operates with an eroticization of vision, a sexualized optical regime. For at the centre of these visual productions stands the voyeuristic exposure of naked corpses, "the body's anatomical nudity."[43] But this nakedness is accentuated by an ideographic sexualization, for the flayed bodies are endowed with either an enormous penis or voluptuous breasts. Even in death, the body is shown to be visually seductive.

The transformation of human corpses into sexual objects received nationwide attention in Germany when the exhibition moved to Hamburg's Erotic Art Museum in 2003. The choice of this site not only points to an enhanced appetite for libidinal signs, but also reveals the installation's pornographic appeal. German media headlines celebrated the "striptease of corpses," the "well-endowed bodies" shown "ultra-nude and uncensored."[44] Not surprisingly, the Erotic Art Museum exhibition coincided with the curator's decision to expand the installation's section on sex and reproduction, and to place pertinent specimens, including "the erect penis of one of the dead, a clitoris, and a foetus in the mother's womb,"[45] into the center of the exhibition. In Germany, death and sex—corpses with sex appeal—have market value.

Misogynist Corporeality

The exhibition's collection of body parts contains numerous healthy organs, including nearly a dozen sets of male genitalia. These plastinated and often elongated or erect penises, complete with scrota, are displayed in separate showcases. By contrast women's reproductive organs, that is, vagina and uterus with ovaries, are exhibited with malignancies, tumors, deformities, and cysts. The Mannheim exhibit in 1997 contained only two female figures: both were pregnant women. In one woman's corpse, the uterus had been cut open to reveal a five-month-old fetus.[46] In the exhibition catalog this female body is mapped by textual markers which, much in the vein of pornographic techniques of accentuating erogenous zones, highlight specific regions of the body. Here, however, the voyeuristic gaze is directed to the female interior—the intestines, the organs of elimination, the uterus and the fetus, a presentation that is suggestive of a preoccupation with female secretions and body products: excrement, urine, placental tissue and fetal substances.

In the exhibition, as in the catalog, the pregnant women are grouped with an assortment of pathologies—malformed embryos, miscarriages, and aborted fetuses: "In a glass case at the center of [this] room, visitors encounter a row of plasticized infant corpses, including a pair of conjoined twins."[47] The subsequent exhibition in Basel, Switzerland, likewise "shows Siamese twins or brain-damaged [infants] together with embryos, fetuses and a pregnant woman's body that has been cut open."[48] In London, the visitor's gaze was similarly directed towards "a pregnant woman reclining demurely in a room lined with deformed fetuses and other 'pregnancy complications.'"[49] Such an iconographic emphasis on female reproduction gone awry—the female body as pathology—is symptomatic of a misogynist conception of womanhood. In this German exhibition about human anatomy, a national body politic is commemorated by museum preservation, and inscribed and implanted into corpses by a medical gaze preoccupied with reproduction and sex. By way of these biological phantasms, culturally specific notions of sexuality are reified as natural or authentic anatomy. This fabrication of biomedical truth is furthermore obfuscated by the aestheticized presentation of the exhibited objects, which circumvents the possibility of critical reflection.

Corpse Art

This exhibition of human bodies is driven by an aesthetic that seeks to commodify corpses as "sculptures"—artistic figures. What visitors supposedly see is the "art of anatomy: the aesthetically instructive presentation of the body's interior."[50] According to media reports, the "remarkable success of the current Body Worlds exhibition" is linked to the fact that "dissected human cadavers are ... presented as works of art."[51] Artistic intervention is a critical aspect of the presentations: "The corpse must be displayed artistically, otherwise it becomes an object of revulsion and obstructs the unemotional gaze":[52] "the aesthetic pose helps to dispel disgust."[53]

The installations are supposed to instruct without eliciting any emotions. The dead, which are "here shown in a highly artistic form of preservation," are "beautiful and educational."[54] "No offensive odors disturb the viewing," because plastination prevents the onset of decomposition in the lifeless bodies.[55] The corpses on display are "dry, hard and scent-free,"[56] "odorless,"[57] "clean," "non-infectious," and "one can touch them ... with one's own hands ... as often as one likes,"[58] "even without gloves."[59] Indeed, the exhibition explicitly extends "an invitation to touch and to look into the interior of the body."[60] Since the individual exhibits are strategically "placed in the very center of the walkway, the tactile contact with the artful dead is in fact palpably provoked."[61]

Such a contrived proximity to the corpse is suggestive of a "democratization of anatomy,"[62] for the exhibited objects—being "resilient, enduring and realistic"[63]—are supposed to offer the visiting public "seemingly direct access" to the corpses.[64] The museum display is designed to facilitate "a sensory awareness of the world of the body through the immediacy of visual contact with the eye and the comprehending touch of the hand."[65] Such a sensualized exhibition undoubtedly reinforces the feeling of a material presence of real and authentic corpses—a presumed intimacy with death.

Of course, this contact with and proximity to death is a staged illusion, a museum effect, because the displayed objects are not stinking corpses but synthetically sanitized bodies: "dry and odorless," "graspable," lacking "decomposition and desiccation so completely that the interior of the body ceases to be an object of revulsion."[66] "The show wants to ... diminish the horror of death."[67] But the installations retreat precisely from this reality of death by exhibiting "living" corpses, normalized and eroticized. The prominent characteristics through which death might be experienced, the sensually apprehensible signs of decay, have been intentionally eliminated from the exhibits. These staged encounters with dead bodies simultaneously supply the necessary emotional distance.

Central to this exhibition is "the aestheticization of corpses. By means of the plastination technique, the exhibits become in large part unrecognizable as real bodies; this art form promotes a forgetting of the fact that these are indeed dead human beings."[68] One "has to repeatedly remind oneself that this is not an artificial panopticon, but true anatomy.... The dissected bodies look ... so deceptively lifelike that they indeed appear to be works of art."[69] These observations are important, for they are suggestive of the fact that the installation effect intentionally negates the proclaimed proximity to death. By means of the "aesthetic dimension of showing,"[70] the corpses are presented in such a way that they appear to be un-dead. The figure of *The Muscle Man*, for instance, "skillfully performs weightlifting experiments."[71] The anatomical exhibition of corpses is "lively, active and posed."[72] The "beauty [of these corporeal forms] displaces our revulsion"[73]; "aesthetics becomes a means for pushing away disgust."[74] Out of the raw material of human remains new bodies have been fashioned — aestheticized, durable, and sensually accessible.

Plastination remakes the corpse into an aesthetic object: with his flesh restored and rendered immortal, *The Skin Man* stands transfixed, focused only on himself. A set of motifs that mark the transfiguration of this corpse — white skin, "classical" Caucasian facial features, a muscular body and heroic pose — suggests a return to a disturbing fascination with fascist masculinity. And presented without reference to the deceased subjects' personal biography, life history, or cause of death, the plastinated specimens perform an inverse symbolic function. For the corpses are aestheticized in such a way as to suppress any evocations of violence, victimhood, or history: the performative success of the specimens is derived from the amputation of feelings and the erasure of memory. Plastination renders the dead bodies "timeless and thereby removes the potential of forgetting or remembering."[75] The displayed objects, which are repeatedly described as "anatomical art works," successfully evade the memory work of history: they possess neither an autobiographical memory nor a historical remembrance. The installation suppresses the commemoration both of the dead subjects' own life histories and of Germany's problematic history of medical experiments, eugenics, and racial hygiene under National Socialism. Indeed, the refurbished corpses are crafted in the tradition of the classic white German body: "eternal human monuments" in a "museum without forgetting, therefore without remembrance, which means without loss, without mourning and, as such, without empathy."[76]

Negations of Humanity

The practice of corpse art is grounded in a series of negations, a negative dialectic that facilitates the public showing of corpses. The bodies are estranged, depersonalized and reified. Dead bodies are morphologically transformed into "living" corpses. Lifeless matter is sculpted into life-like poses. Mortality is denied by an emphasis on the "resurrection" of the dead body. Temporality is negated by means of synthetic preservatives. The inevitable deterioration of the body is pre-empted: the corpses remain forever "frozen" somewhere "between death and decay,"[77] and hence come into view as an eternally enduring monumental body-architecture.

The processed corpses have been rendered anonymous. The dead subject, the very identity of this formerly living human being, has been erased from the displays: "the historical-biographical construct of the person of the deceased [is] excised from the plastinated objects";[78] moreover, the "age and cause of death are generally withheld in order to avoid a 'personification' of the specimens."[79] Every remnant of subjectivity is destroyed, for the dead bodies are transfigured beyond recognition: even their facial features have been changed. The exhibition shows "peeled, perforated, opened and halved ... dissected bodies, a 'corpse inside the corpse', as it were,"[80] which stand there frozen in time, without personal identity or life history, robbed of its humanity.

But these dead bodies, through the negative labor of technological intervention, are not merely estranged and transfigured, but also objectified. The "cut up, dismembered and recombined three-dimensional 'living corpses'"[81] are no longer recognizable as human individuals. This process of dehumanization has manifest legal implications. For these subjectivity-deprived bodies can be stripped of their civil rights. The insistence on anonymity, the elimination of identity, scientific objectification, and aesthetic figuration collude to neutralize any potential judicial concerns. In most German states "burial regulations generally prohibit the public display of corpses.... But the Mannheim judiciary [in the state of Baden-Wurttemberg] determined that plastinated bodies are not 'corpses in the eyes of the law.'"[82] The exhibited specimens are not human bodies but lifeless matter, material objects. Thus legally defined as a "thing," the plastinated corpse loses its humanity. The exhibition also removes all signs of personhood: the dead bodies are dehistoricized, stripped of any social markers or biographical indicators, and therefore also devoid of remembrance. The corpse "ceases to be the object of reverence and human compassion."[83] "[T]he work of mourning is extinguished ... the sense of emotional attachment to the deceased [has] come to an end."[84] On display are dehumanized "subjectless bodies,"[85] stripped of any affective attention. Thus rendered emotionless, the German body museum "takes its leave from the culture of conscience," from the "drama of the culture of shame."[86]

Recycling Death

The dead are redesigned to become lifeless matter: they are described as medical plastinates, anatomical specimens, museum exhibits, and juridical objects— as thing-like figures. Legally dispossessed of the rights of personhood, they are not human subjects, and thus not even corpses. By means of this negation of human subjectivity, the exhibition creates "phantasms of disposition about the biomass," which "reduces the human subject ... to a functioning apparatus"[87]: the human being "becomes a body ... a device for study."[88] The anatomically bared bodies are variously dissected to reveal specific biological functions—the nervous system, the digestive tract, the circulatory system, or the respiratory apparatus. What is shown is the "human being as a construct."[89] Instrumentalized and objectified, these "nameless figures"[90] are presented with "a decisive purpose" as "objects of knowledge."[91] Within this operational scheme, in which the body is only recognizable as an apparatus, a system, or a tract, the dead are posed as task-oriented action figures: as "swordsman," "spear thrower," "swimmer," "runner," "dancer," "chess player," "rider," and "organ presenter." The installations attempt to create a functional death.

Von Hagens "wants to extract as much 'didactic profit' from the dead as possible,"[92] to make the dead "serve an instrumental purpose."[93] In this anatomical museum, the dead can be "disposed of" in "a meaningful way,"[94] because they are "used for educational ends."[95] But this "post-mortem exhibitionism," as the media termed it[96] operates rather more like a posthumous recycling procedure, a commercial enterprise based on principles that we should recognize from the Nazi era: the attempts to profit from corpses, the recycling of the dead and their body parts, the medical exploits with racialized human material. According to von Hagens, "the plastinated specimens ... are no longer objects of mourning, but instructional objects—with the intended purpose to educate ... 'to serve a useful purpose after death.'"[97] Through medical intervention, he asserts, the dead body undergoes "a shift in value from a useless corpse to a useful, aesthetically instructive ... plastinated specimen."[98] This rational economy, which simultaneously devalues and reclaims value by recycling, and which wants to extract surplus value from the bodies of the dead, is rarely acknowledged in public discussions. And yet it is part of a cultural logic, a capitalist mode of thinking, whereby historical consciousness is repressed in favor of an operational instrumentality: not only the living but also the dead must be usefully productive. According to von Hagens, emotions are an "intrusive" factor, a "digressive" obstacle, in this enterprise: "mourning would interfere with learning."[99] The exhibition stages a dehistoricized morphological landscape, where dead bodies are shown without commemoration and without compassion.

Unlike the mummy, the embalmed body or the venerated relic, the plastinated corpses are supposed to be self-referential or auto-iconic — emptied of meaning, emptied of symbolic content, and devoid of any emotional investment. The presentation of the corpse in this German museum exhibition constitutes the antithesis of the performative efficacy of "the political life of dead bodies," as Katherine Verdery[100] has phrased it: it is relieved of any representational, evocative power with regard to memory and history — death without emotional significance and without political cosmology. Indeed, by legally annulling the humanity of the corpses, the German state dissociates itself from the exhibition and thus denies moral accountability. In this human museum, contrary to Verdery's argument[101], dead bodies are *not* "a site of political profit." The displays of bodies, as I have shown here, have been invested with different meanings.

The Limits of Empathy

Inhumanity, techno-rational regimes of death, and the "banality of evil"[102] succinctly describe the governing principles of the modus operandi of modernity in the twentieth century and, moreover, reveal the formation of a new kind of subjectivity — the increasing atrophy of empathy. Emotional anesthesia, as I have suggested, is integral to a cultural apparatus that feeds on the labor of the negative. The Body Worlds exhibition offers a particularly characteristic example of this form of productivity. Although processes of affect negation may be symptomatic of global capitalism, they are simultaneously embedded within specific histories, discourses, and meanings. In Germany, emotional regimes are inevitably part of post–Holocaust culture, shaped by conflicted memories of catastrophic nationalism. The contact with mutilated corpses, previously evocative of war and genocide, has been rendered safe in the cocooned space of the museum.

What do we make of this "dead body politics" in unified Germany? The parade of corpses, exhibited in a museum, which brings them "into the realm of the timeless or sacred, like an icon,"[103] links a transformed nation to a seemingly frozen corporeal landscape. In this context death may be a metaphor for the end of an era, but plastinated corpses— as symbolic vehicles devoid of historical memory — also point to a renewed resistance to re-evaluating the past. "What is it about a corpse," asks Verdery, "that seems to invite its use in politics, especially in moments of major transformation?"[104] The body's materiality, confirmed by sight and touch, is critical to its symbolic efficacy. Strategically located in the conventional memory site of the museum, the plastinated bodies speak to cultural practices of unmaking remembrance. In this museum of immortality, German visitors see death without mourning. It is a corporeal landscape, staged without historical reflexivity, and thus also without the engagement of feelings. The dead, robbed of their humanity, display their seemingly un-dead bodies with a scientific objectivity that undoes and negates the museum's task of memory production. In Germany, the public intimacy with corpses is indicative of a psychic economy that, with the return of the repressed, simultaneously moves in an eroticized void of remembering and forgetting.

Notes

1. Halbwachs, *The Collective Memory*; Nora, *Les lieux de mémoire*; Le Goff, *History and Memory*.
2. Korff, "Reflections on the Museum."
3. Baudrillard, *Die fatalen Strategien*.
4. Huyssen, *Present Pasts*, 11–24.
5. de Certeau, *The Writing of History*.
6. Linke, "The Anthropology."
7. Linke, "Volks—"
8. Kleinman and Kleinman, "The Appeal."
9. Schenda, *Gut bei Leibe*.
10. Clewing, "Schneller," 16.
11. Landesmuseum, *Körperwelten*.
12. Roth, "Body counts," 50.
13. van Dijck, "Bodyworlds"; Walter, "Plastination"; Walter, "Body Worlds"; Csordas, "Computerized."
14. Herscovitch, "Rest in Plastic," 828.
15. Clewing, "Schneller," 16.
16. Roth, 51, 52.
17. von Hagens, "Der plastinierte," 214.
18. Becker, "Zugriff," 15.
19. Nissen, "Anatomie," 36.
20. Vorpahl, "Der Tank," 8.
21. Becker, 15.
22. Budde, "Der sezierte," 11.
23. Schmitz, "...was die Welt," 8.
24. Korff, "Reflections."
25. Nissen, 36.
26. Anon, "Köperwelten' waren," 33.
27. Kirschenblatt-Gimblett, *Destination Culture*, 35.
28. Ibid., 36.
29. von Hagens, "Der plastinierte," 214.
30. in Roth, 52.
31. Anon, "Stachel," 214.
32. Herscovitch, 828.
33. Andrews, "Anatomy," 1.
34. Zoschke, "Qual des letzten."
35. Roth, 52.
36. von Hagens, "Der plastinierte," 203–04.
37. Schmitz, 8.
38. von Hagens, "On Gruesome," 260.
39. Becker, 15.
40. Anon, "Mut zur," 184.
41. von Hagens, "On Gruesome," 214.
42. Horkheimer and Adorno, "Interesse am Körper," 277.
43. Whalley, "Der menschliche,"161.
44. Assheuer, "Die Olympiade."
45. Ibid.
46. Landesmuseum, 136, 138, 140–1.
47. Andrews, 4.
48. Becker, 15.
49. Herscovitch, 828.
50. von Hagens, "Der plastinierte," 214, 217.
51. Smith, "Nota bene," 829.
52. da Fornseca, "Wachsfigur—," 50.
53. von Hagens, "Der plastinierte," 205.

54. Kriz, "Einführung," 9–10.
55. von Hagens, "Der plastinierte," 205.
56. Vorpahl, 8.
57. Scheytt, "In Silikon," 1–2.
58. Roth, 50; Clewing, "Hohe Kunst," 3; Zoschke, 12.
59. Rothschild, "Das nennt man," 3.
60. Schmitz, 8.
61. Roth, 51.
62. Anon, "Körperwelten," 33.
63. Zoschke, 12.
64. Becker, 15.
65. Bauer, "Anatomie," 199.
66. von Hagens, "Der plastinierte," 204.
67. Kriz, 9.
68. Emmrich, "Sich selbst," 6.
69. Schmidtz.
70. Clewing, "Vier gegen," 30.
71. Roth, 51.
72 von Hagens, "Der plastinierte," 204.
73. Eggebrecht, "Mehr als," 1.
74. Reimer, "Nächtliche," 2.
75. da Fonseca, 58.
76. Ibid., 56, 60.
77. von Hagens, "Der plastinierte," 201.
78. Bauer, 197.
79. von Hagens, "Der plastinierte," 201.
80. Roth, 51–52.
81. Ibid., 52.
82. Eberhardt, "Leichen im," 18.
83. Vorpahl, 8.
84. von Hagens, "Der plastinierte," 207.
85. da Fonseca, 66.
86. Lethen, *Verhaltenslehren*, 6–7.
87. Becker, 15.
88. Budde, 11.
89. Ibid.
90. da Fonseca, 44.
91. von Hagens, "Der plastinierte," 214.
92. Anon, "Über das," 22.
93. Eberhardt, 18.
94. Vorpahl, 8.
95. von Hagens, "Der plastinierte," 207.
96. Rothschild, "Das nennt," 3.
97. in Eberhardt, 18.
98. von Hagens, "Der plastinierte," 215.
99. Ibid., 203–7, 214.
100. Verdery, *The Political Lives.*
101. Ibid., 33.
102. Arendt, *Eichmann.*
103. Verdery, 5.
104. Ibid., 27.

Bibliography

Andrews, Edmund L. "Anatomy on display." *New York Times*, January 7, 1998.
Arendt, Hannah. *Eichmann in Jerusalem*. New York: Penguin Books, 1994.

Assheuer, Thomas. "Die Olympiade der Leichen." *Die Zeit 35*, August 21, 2003 http://www.zeit.de/2003/35/K_9arperwelten (accessed February 17, 2008).

Baudrillard, Jean. *Die fatalen Strategien*. Berlin: Matthes & Seitz, 1991.

Bauer, Axel W. "Anatomie und Öffentlichkeit—ein bioethischer Problemfall?" In Landesmuseum für Technik und Arbeit Mannheim and Institut für Plastination, *Körperwelten*, 195–200.

Becker, Jochen."Zugriff auf die Biomasse." *tageszeitung*, October 11, 1999.

Budde, Kai. "Der sezierte Tote." In Landesmuseum für Technik und Arbeit Mannheim and Institut für Plastination, *Körperwelten*, 11–28.

Clewing, Ulrich. "Schneller Altern im Ideenkorsett." *tageszeitung*, September 7, 1998.

_____."Hohe Kunst oder 'Ärgernis,'" *tageszeitung*, October 8, 1998.

_____."Vier gegen Keinen." *tageszeitung*, October 10, 1998.

Csordas, Thomas J. "Computerized Cadavers." In *Biotechnology, Culture, and the Body*, edited by Paul Brodwin, 173–92. Bloomington: Indiana University Press, 2000.

da Fonseca, Liselotte Hermes. "Wachsfigur—Mensch—Plastinat." *Deutsche Vierteljahresschrift für Literaturwissenschaft und Geistesgeschichte* 73, no.1 (1999): 43–68.

de Certeau, Michel. *The Writing of History*. New York: Columbia University Press, 1988.

Eberhardt, Johanna. "Leichen im Sinne des Gesetzes?" *Frankfurter Rundschau*, December 17, 1997.

Eggebrecht, Harald. "Mehr als nur Haut und Knochen." *Süddeutsche Zeitung*, January 17, 1998.

Emmrich, Michael. "Sich selbst begreifen." *Frankfurter Rundschau*, February 28, 1998.

Halbwachs, Maurice. *The Collective Memory*. New York: Harper & Row, 1980.

Herscovitch, Penny. "Rest in Plastic." *Science* 299 (2003): 828.

Horkheimer, Max and Adorno, Theodor W. "Interesse am Körper." In *Dialektik der_Aufklärung*, edited by Max Horkeimer and Theodor W. Adorno, 276–81. Frankfurt/M: Suhrkamp, 1981.

Huyssen, Andreas. *Present Pasts*. Stanford: Stanford University Press, 2003.

Kirshenblatt-Gimblett, Barbara. *Destination Culture*. Berkeley: University of California Press, 1998.

Kleinman, Arthur, and Joan Kleinman. "The Appeal of Experience, the Dismay of Images." In *Social Suffering*, edited by Arthur Kleinman, Veena Das, and Margaret Lock, 1–24. Berkeley: University of California Press, 1997.

Korff, Gottfried. "Reflections on the Museum." *Journal of Folklore Research* 36, no.2/3 (1999): 267–270.

_____. *Museums-Dinge*. Cologne: Böhlau, 2002.

"'Körperwelten' waren Magnet fürs Publikum." *Frankfurter Rundschau*, March 3, 1998.

Kriz, Wilhelm. "Einführung." In Landesmuseum für Technik und Arbeit Mannheim and Institut für Plastination, *Körperwelten*, 9–10.

Landesmuseum für Technik und Arbeit and Institut für Plastination Heidelberg, eds. *Körperwelten*. Heidelberg: Institut für Plastination, 1997.

Le Goff, Jacques. *History and Memory*. New York: Columbia University Press, 1992.

Lethen, Helmut. *Verhaltenslehren der Kälte*. Frankfurt/M: Suhrkamp, 1994.

Linke, Uli. "The Anthropology of Collective Memory." *International Encyclopedia of the Behavioral and Social Sciences* 3, no. 1 (2001): 2219–23.

_____. "Volks—Körper—Kunde." In *Unterwelten der Kultur*, edited by Thomas Hauschild and Bernd JürgenWarneken, 65–94. Cologne: Böhlau, 2003.

"Mut zur Todesnähe." *Der Spiegel* 36, August 31, 1998, 182–85.

Nissen, Klaus. "Anatomie nach Mitternacht." *Frankfurter Rundschau*, February 27, 1998.

Nora, Pierre. *Les lieux de mémoire*. Paris: Gallimard, 1976.

Reimer, Ulf. "Nächtliche Wallfahrt zu den Toten." *Süddeutsche Zeitung*, March 2, 1998.

Roth, Jürgen. "Body counts." *Konkret* 4 (1998): 50–52.

Rothschild, Markus. "Das nennt man 'postmortalen Exhibitionismus.'" *tageszeitung*, October 8, 1998, 3.

Schenda, Rudolf. *Gut bei Leibe*. Munich: C. H. Beck, 1998.

Scheytt, Stefan. 1997 "In Silikon und Ewigkeit." *Die Zeit*, August 14, 1997, 1–3.

Schmitz, Helmut. "...was die Welt im Innersten zusammenhält." *Frankfurter Rundschau*, November 12, 1997.

Smith, Orla. "Nota bene: Anatomy." *Science* 299 (2003): 829.
"Stachel im Fleisch." *Der Spiegel* 41, October 6, 1997, 212–4.
"Über das Sterben." *Frankfurter Rundschau*, November 22, 1997.
van Dijck, José. "Bodyworlds." *Configurations* 91 (2001): 189–216.
Verdery, Katherine. *The Political Lives of Dead Bodies*. New York: Columbia University Press, 1991.
von Hagens, Gunther. "Der plastinierte Mensch." In Landesmuseum für Technik und Arbeit Mannheim and Institut für Plastination Heidelberg, *Körperwelten*, 201–216.
_____. "On Gruesome Corpses, Gestalt Plastinates and Mandatory Interment." In *Gunther von Hagens' Body Worlds*, edited by Gunther von Hagens and Angelina Whalley, 260–282. London: Institut für Plastination, 2004.
_____, and Angelina Whalley, eds. *Gunther von Hagens' Body Worlds: The Anatomical Exhibition of Real Human Bodies*. London: Institut für Plastination, 2004.
Vorpahl, Annette. "Der Tank für den Elefanten steht bereit." *Frankfurter Rundschau*, October 25, 1997.
Walter, Tony. "Plastination for Display." *Journal of the Royal Anthropological Institute* 10 (2004): 603–627.
_____. "Body Worlds." *Sociology of Health & Illness* 26 (2004): 464–88.
Whalley, Angelina. "Der menschliche Körper." In Landesmuseum für Technik und Arbeit Mannheim and Institut für Plastination, *Körperwelten*, 49–194.
Zoschke, Barbara. "Qual des letzten Willens," *tageszeitung*, April 3, 1994.

Section Four

Disharmonious Bodily Openings

Forced Impregnation and Masculinist Utopia

T. Christine Jespersen and Alicita Rodríguez

Gunther von Hagens' Body Worlds exhibits have drawn more than 20 million viewers worldwide, most wishing to see the whole body specimens that the anatomist labels gestalt plastinates.[1] While von Hagens' plastination technology continues to attract medical professionals and lay persons alike, his work demands a critical focus, because his human specimens literally and symbolically fix the body. Further, because he uses both male and female cadavers to create his anatomical specimens, von Hagens also fixes gender, defining gender in essentialist fashion. Herein lies the danger: these fixed anatomies become equated with the natural; consequently, science appears transparent. This is the human body, the exhibited cadavers seem to say. But the natural is, in fact, a cultural construct that has been used to justify countless positions, including scientific and anthropological racism, as well as sexism. The body becomes particularly problematic because an anatomical specimen is the nexus of the natural. Lisa Jean Moore and Adele E. Clarke tackle this issue in their study of representations of the clitoris, finding that "[a]natomy is one of the key sites for the production and maintenance of sex and gender."[2] Neuroscientist Steven Rose concludes that the human body is impossible to define absolutely: "In a world which is understood in terms of process rather than object, the joints into which we carve nature depend on our ultimate purposes...."[3] In his discussions of race, Stuart Hall also explores how the concept of the natural supports cultural claims.[4] Von Hagens asserts that Body Worlds demystifies the human body and educates the public in a spirit of democracy. If Body Worlds is didactic, we should be especially wary: What exactly are these plastinates teaching us? Dr. von Hagens' insistence that his bodies are authentic—more authentic than the body before plastination—further problematizes his exhibits.

Von Hagens casts Body Worlds as authentic and universal, yet that "world" is a cultural construction invented as a new Eden where birth is usurped and gender meaning is fixed. The process of plastination and the organization of the exhibit literally and symbolically paralyze and fragment the human body, replacing the natural world with a dream world of immortality and rigidity. The rejection of the fluidity of the pre-plastinated physical body with all of its water, blood, and guts and the installation of a secured cultural one have significant implications regarding gender. Fixed bodies, within a re-created world, force essentialist definitions of body and gender.

In much of his writing, von Hagens self-consciously stages himself as the creator of a new world. He refers to himself as a designer, an inventor, an engineer, an architect, a chess player, and a sculptor. Eerily suggestive of a macabre Garden of Eden, von Hagens invents a utopia where de-animation and death spawn immortality. One of the "design principles" of Body Worlds is the placement of plastinates in "a lifelike and attractive setting, e.g., one enlivened with plants and objects from daily life"; for example, some plastinates "have been placed in a garden setting complete with a brook."[5] Further evocative of a redesigned garden of innocence, he notes the extreme nakedness of plastinated bodies, so naked that they are without skin.[6] Von Hagens' preoccupation with the staging of his denuded bodies emphasizes that he must not only create new beings, but also devise a new and better world for them. An advantage of plastination, von Hagens enthuses, is the achievement of a "certain immortality" that defeats the "'memento mori' of the Church and the teaching on 'dust to dust.'"[7] The world, then, that von Hagens engineers is one that defeats death and the life cycle by replacing it with plastinated corpses molded and fixed to von Hagens' specifications. He explains that the word plastinate is his "own invention" and derives from the Greek "plassein" meaning "to mold, form."[8] Discussions of positioning, transforming, creating, and inventing — placing body parts into what he calls gestalt forms — dominate his writing.

Although one might object to Body Worlds on many grounds, this paper focuses on the gendered implications of von Hagens' plastinated Eden. Von Hagens professes to have invented the plastination process; the book *Gunther von Hagens' Body Worlds: The Anatomical Exhibition of Real Human Bodies* reproduces the United States patent rights for any disbelievers. The patented process, which legally secures financial gains and symbolically renders von Hagens the sanctioned and only inventor, involves the replacement of bodily fluids "by acetone through diffusion" and then the substitution of the acetone with reactive plastics.[9] When the process is completed, bodies are initially pliable. Von Hagens explains, "Hardening cannot take place immediately after the specimen has been removed from the silicon tank because at that point it is initially a crumpled heap of matter. In this situation, we are virtually compelled to shape it in some way."[10] Notably, von Hagens terms the phase in which plastic is infused into the body "forced vacuum impregnation."[11] The German "imprägnieren" is often used in scientific procedures, yet, as in English, it means to impregnate and carries similar connotations.* Von Hagens himself emphasizes the link to human pregnancy: "The decisive trick, with which liquid plastic can be infused into the last cells, is forced vacuum impregnation. Just as babies use a partial vacuum to suck the milk out of their mothers' breasts, acetone is sucked from the specimen in the vacuum. In this way, the vacuum created in the specimen keeps a steady supply of plastic flowing into the tissue."[12] This bizarre description at once connotes rape, abortion, and a feminized and fragmented conception of the body—where breasts become the synecdoche for the abject body yet to be "animated" into a frozen gestalt gesture. Plastic replaces maternal milk, so that the body, rendered a "crumpled heap of matter," might be shaped and hardened. The violence of the procedure is implicit in the language as is the usurpation of the birthing process from women.

Von Hagens makes clear his preference for plastic to bodily fluids. He refers often to plastic as "aesthetic" and "clean" and to the plastination process as lending corpses "a fresh, plastic appearance and texture, [that] makes them beautiful."[13] To von Hagens the sterile aestheticism of the inert plastic supplants the mess of mother's milk and unruliness of *homo sapien* sperm in order to "birth" exquisite, synthetic, and static corpses. Von Hagens' sanitized and controlled process replaces sex between actual human beings. He extols the virtues of "forced impregnation with plastic in a vacuum" as follows:

1. Impregnation can be done with a number of liquid, polymerizeable plastics
2. By regulating the vacuum, the speed of impregnation can be adapted to the type of specimen, its thickness and the viscosity of the plastics used
3. Impregnation can be used to minimize shrinkage at low temperatures, e.g., at -25°C.[14]

Polymerizeable plastics determine the "look and feel of the finished specimen."[15] Thus, careful selection of material controls appearance and touch of the final progeny. Preventing any unplanned "pregnancy" or unforeseen outcome, plastic—regulated for thickness and viscosity—replaces sperm. Even speed is constrained. Perhaps the measure prevents [premature] ejaculation? Although von Hagens obviously refers to shrinkage of tissues in the corpse, the double entendre of the third "advantage" suggests prevention of masculine vulnerability.

In fact, von Hagens' Body Worlds carefully excises any emotion or emotional contact. In several different places, von Hagens explains the dangers of emotional attachment to the "specimens." For example, he defends anonymity on the following grounds: "Ensuring anonymity," he writes, "is important for distancing the body from its plastinated counterpart, as it is the only sure way of ending the sense of reverence surrounding that body, i.e., the sense of personal and emotional attachment to the deceased."[16] This emotional detachment

**Thanks to Vanessa Manhire for the translation.*

has at least two effects. In the first place, once the body passes into von Hagens' hands, it ceases to be a human being and becomes an object of his creation, "mere raw material."[17] Although he seeks to maintain the individual imprint of the body below the surface of the skin, he sees himself as an inventor who improves upon the original. He argues, for example, that "the design efforts of a plastinator actually affirm his intention to upgrade the value of the human body."[18] In the context of describing how he chooses body positions for the plastinates, von Hagens asks, "What is the ethical-moral difference between the glass eye of a gestalt plastinate and the glass eye used in a reconstructive facial operation or between a gesture of movement in a gestalt plastinate and a breast enlargement in a living person?"[19] Contending that plastination is both ethical and moral, von Hagens equates plastination with reconstructive and plastic surgery in living persons. He seems to regard his work as providing for deficiencies and improving upon the human form — the human without the eye and the woman with small breasts. The comparison of movement to breast enlargement is especially disturbing when one considers the comparative importance in daily life between movement and breast enhancement. Between exhibits, in at least one case, he actually replaced one female body with another who had "conspicuously enlarged" breasts.[20] In another celebration of plastination as an improvement over real bodies, von Hagens writes, "bodies can ... undergo a change in meaning: from a grisly unknown quantity to an intimate main attraction of creation."[21]

In the second place, the removal of emotion in opposition to personal detachment helps to support a larger, fixed dichotomous structure imbued in the philosophy, process, and presentation of the plastinated bodies. In reference to Donna Haraway's "Cyborg Manifesto," Megan Stern argues, "Read as cyborgs, the visible humans and plastinates have the potential to destabilize the traditional representation of patient as passive, female, physical object and doctor as active, male, cerebral subject."[22] In Haraway's well-known theorization, the science-fiction cyborg can help us to conceptualize a post-gender, post-race, post-class world by disrupting a knowledge system based upon binary opposition. The cyborg, she contends, "signals three crucial boundary breakdowns" human and animal, human and machine, and physical and non-physical.[23] Stern concludes, however, that *Body Worlds* ultimately fails to "challenge the polarized understanding of gender and knowledge."[24] Body Worlds, however, does more than that. It determinedly cements and defends them. We see this in von Hagens' exhibitions of the female bodies, his taxonomy of corpses, and the physical layout of the museum.

As others have noted, for the most part representations of female bodies in the exhibit are conventional. When women are present at all, they are figured in a couple of ways. They are sexualized and/or pregnant —comedian Margaret Cho ruefully queries, "Why are men science and women pornography?"[25] — or they are exhibited as performing sanctioned female activities such as ice-skating, dancing, and yoga. The female plastinates are displayed in opposition to their male counterparts who are swordsmen, ski jumpers, skate boarders, bicyclists, and players of soccer, basketball, and chess. Von Hagens' fascination with male hardness and musculature translates into a taxonomic dualism that abhors softness, fluidity, wetness, movement, and emotion. He goes so far as to develop a taxonomic graph delineating the quality of corpses. On the one hand are "cadavers" and "wet cadavers." Cadavers, the least desirable, are "decomposing," "desiccating," "to be mourned," and "decomposing body parts." Wet cadavers are also decomposing and desiccating. They include "bog bodies, glacier corpses," and "dissections preserved in aqueous solution."[26] On the other hand are the superior "dry cadavers" with none of these disgusting qualities. They include mummies, skeletons, and, of course, gestalt plastinates.

Von Hagens continuously refers to these plastinates as authentic. In a typical example, he writes, "Beyond their didactic qualities, plastinated specimens also radiate a certain fascination, based primarily on their authenticity. Plastination stops decomposition and dehydration so completely that the insides of bodies cease to be objects of revulsion. Observers are not bothered by any kind of offensive odors."[27] Cast in opposition to the clean, dry, and fascinating plastinates, real bodies are characterized as decomposing, wet, revolting, and radiating offensive odors. Authenticity is linked by juxtaposition to the arrest of the natural processes of decomposition and to the disappearance of fluidity, which makes the bodies "colorful, not smelly."[28] In a sleight of hand, then, not only are the plastinates rendered superior to bodies, but are also made authentic through plastination.

The implications of rendering plastinated dead bodies more authentic than living human beings, of course, extends far beyond gender. However, to return to the focus of the paper, we move now to a close reading of the fetal development room in the Body Worlds 2 exhibit at the Denver Museum of Nature & Science. The presentation of this room magnifies the problems of authenticity and binary structure apparent elsewhere in Body Worlds.

Before Body Worlds 2 arrived at the Denver Museum of Nature and Science, the organizers of the exhibition decided to house the specimens of a pregnant woman and fetuses at various stages of development, including deformed fetuses, in a room separated from the rest of the exhibition by drapery-covered panels— a partition likened to a "dark curtain of shame"[29] and a "shielding fabric tunnel."[30] Partially, this decision stemmed from prior vandalism: in London, one man first covered the whole-body plastinate of a pregnant woman, then spilled paint on the floor in protest; and in California, two women stole one of the exhibit's fetuses. The Denver Museum of Nature and Science did, however, emphasize the fact that Body Worlds 2 contained cadavers of a pregnant woman and fetuses by noting it on their website, as well as on their telephone recording that provided information about the exhibit. Additionally, the Denver Museum of Nature and Science chose to play classical music (the soundtrack provided by the Body Worlds exhibit) in what we will refer to as the fetal room, even though there was no audible music in the rest of the exhibit. The lighting in the fetal room was noticeably dimmer than in the main exhibition area, which was extremely well-lit.

In the midst of the loud main area of von Hagens' new Eden, peopled almost exclusively by male whole-body plastinates in athletic poses, a black wall loomed with only a small plaque describing the optional exhibition space. The fact that the fetal room presents itself as an afterthought (coming at the end of the whole-body specimens) and as an option is troubling: it implies that the female body and its control over the birthing process are secondary to the dominant male figure, especially as fixed by the creator, Dr. von Hagens. All of the gestalt plastinates are signed by him, an act that "contrary to the claims of the surrounding literature [von Hagens' book]" erases everyone's individualism save his.[31] Some people who visited the exhibit failed to see the sign at all, furthering the idea that the fetal room was secret; this idea is particularly intriguing when compared to the male protestor's comment that the womb "is a sacred place that doesn't deserve to be looked at."[32] Appearing after the animated male specimens, the fetal room stands in opposition to the lighter, brighter, more engaging, and ultimately preferential world designed by von Hagens. The female body and vaginal birth is the dirty, wet alternative to the clean, dry, odorless beings of Dr. von Hagens. The fetal room also contains deformed fetuses, which creates an association between the pregnant woman and abnormality.

And while no explanation regarding their lives or deaths accompanied any of the male plastinates, the pregnant woman is plagued by her very own informational plaque, which reads more like an indictment: the sign informs us that she had lung cancer and that a fatal disease killed her and her unborn child. Of course, by not specifying the mysterious disease, the sign renders her guilty. Even without the lung cancer, the explanation of the pregnant woman's circumstances evidences von Hagens' gendered binaries: why does the woman's life require an examination that the men's does not? That we must question the woman's existence while ignoring the men's confirms the sexist streak that the history of anatomy showcases—the man is normative, the woman other. The assumed or default body is male.

The first sign that the female body is different, if not abnormal, is literal: the exhibition sign that warns us to enter at our own peril. Positing the female body as other continues with the next sign, which informs us of the pregnant woman's lung cancer. We simultaneously receive the second sign, the black drapes. Once inside the fetal room, deep garnet drapery panels interrupt the black drapes, the one and only reminder of blood in an exhibit of cadavers whose preservation necessitates the extraction of blood and other fluids. All of the drapes are pleated, an effect that emphasizes softness and implies liquidity and flow. These characteristics exist in stark contrast to the hard, muscular male specimens that von Hagens favors. Sociologist Tony Walter argues that von Hagens continues the anatomical traditional of displaying "the female as a lesser form of the male" by virtually excluding female whole-body plastinates due to their inferior musculature.[33] Given the association of women with the domestic sphere, the choice of drapery in the fetal room cannot be underestimated, as it recalls sewing, a stereotypically gendered act. Additionally, the pleated fabric recalls the female womb, linking the only free-standing whole-body plastinate of a woman with reproduction, "implicitly defining the female as a reproductive machine."[34]

The pregnant woman is encased in glass, just like the fetuses, likening her to the exhibit's other anatomical specimens shown under glass—notably body parts and organs that admittedly "come primarily from the female body donors"[35]—and thus objectifying her as a piece of meat. Unlike the male specimens who are free, the pregnant woman appears trapped, like a caged zoo animal. She exists for our gaze; she "is everyone's property in a way that a male body could never be...."[36] Like the fetuses, she nestles on black fabric, which, according to Walter, is "a sign of mourning."[37] The dim lighting in the fetal room is another signal that the female body and its reproductive capability should be kept in the dark. The choice of the museum to play classical music in this room alone adds an uncanny aspect. Should the music relax the pregnant woman, assuaging her guilt for murdering her baby? Is it intended to aid the development of the fetuses, as research has shown to link classical music with growth? Or should it just pacify the viewers, who should be so horrified by the presence of the female form?

The combination of the many staging decisions in the fetal room clearly affected the exhibit's visitors, who abandoned their idle chat and awesome wonder to adopt whispers if not silence. The overall effect was apologetic, as if von Hagens needed to showcase the inferiority of women and progenitive birth—in opposition to his utopian creation. We see how he establishes this dichotomy in his own writing: "Both sides of the body, the attractive and the less attractive, are real—here a perfect, strapping body, there an old and feeble body or even a mutilated one ... here aesthetically posed plastinates, there severely deformed fetuses."[38] The only alternatives to a hard male body seem to be the "feeble" body of women or the "deformed" bodies of fetuses. In fact, there is no room for healthy fetuses at all, due to von Hagens' usurpation of the birthing process. By fixing pregnancy and fetal development, von Hagens unwittingly represents abortion.

Paralysis and fragmentation of movement dominate the fetal room. This is perhaps best exemplified by the embryos floating in precisely labeled, separate glass cylinders that represent the stages of development. But fetal development is a continuous and fluid process—a cycle that von Hagens insists on halting. The male plastinates are virtually defined by their movement, whereas the women and fetuses must be stopped. Women can be shown pregnant; fetuses can be shown; but nowhere are there babies. One can surmise that if pregnant women who died were willing to donate their bodies, including their unborn babies, to plastination, then women with stillborn babies would also be willing to donate their children's bodies. However, neither babies nor children born of women exist in von Hagens' world. He creates an either/or distinction between the healthy, clean, plastinated male bodies, which pose dynamically in his created world, and the sickly, dirty female bodies and fetuses (here read as the products of womanhood), which remain frozen in aborted pregnancies. Von Hagens' choice to feature female bodies reproductively suggests "a preoccupation with female secretions and body products: excrement, urine, placental tissue and foetal substances"[39]—all the wetness of the undesirable cadaver as opposed to the authentic plastinate. Further, the pregnant woman is also housed with the deformed, including "hydrocephalic and conjoined fetuses."[40] The only outcome of female reproduction seems to be abnormality—"deformed fetuses that in no way have any aesthetic appeal" as von Hagens states.[41] Anthropologist Uli Linke rightly asserts, "Such an iconographic emphasis on female reproduction gone awry—the female body as pathology—is symptomatic of a misogynist conception of womanhood."[42]

Ultimately, while the whole-body male peoples a lively and energetic utopia where plastination has "stopp[ed] its individual time" and "kill[ed] death,"[43] the whole-body female exists only in death. She and all the life that she represents must be disassembled and dispersed into discrete and manageable units, or confined and hidden away in a space of arrested birth. All this in order to make room for the masculinist dream of controlled animation, excised of the fluid and messy dynamism of life.

Notes

1. Institute for Plastination, Gunther von Hagens'.
2. Moore and Clarke, "Cliteral Conventions," 2.
3. Rose, *Lifelines*, 42.
4. Hall, "Ethnicity" and Jally, *Race: The Floating Signifier.*
5. von Hagens, "On Gruesome," 266.
6. Ibid., 269.
7. Ibid., 272.
8. von Hagens, "Anatomy," 27–28.
9. Ibid., 21.
10. von Hagens, "On Gruesome," 261.
11. von Hagens, "Anatomy," 21.
12. Ibid.
13. von Hagens, "Why," 2.
14. von Hagens, "Anatomy," 22.
15. Institute for Plastination, "The Plastination Process."
16. von Hagens, "Anatomy," 31.
17. Linke, "Touching," 15.
18. von Hagens, "On Gruesome," 262.
19. Ibid., 267.

20. Linke, 17.
21. von Hagens, "On Gruesome," 269.
22. Stern, "Dystopian," 75.
23. Haraway, *Simians,* 151–53.
24. Stern, 76.
25. Cho, "Body Worlds."
26. von Hagens, "On Gruesome," 277.
27. von Hagens, "Anatomy," 20.
28. von Hagens, "On Gruesome," 24.
29. Allen, "Body Ethics."
30. "Body Language."
31. Johnson, "My Body," 76.
32. Claire, "Paint Protest."
33. Walter, "Body Worlds," 483.
34. Walter, "Plastination," 620.
35. Institute for Plastination, "Questions & Answers."
36. "Body Language."
37. Walter, "Plastination," 620.
38. von Hagens, "On Gruesome," 268.
39. Linke, 17.
40. Johnson, 79.
41. von Hagens, "On Gruesome," 268.
42. Linke, 17.
43. Hirschauer, "Animated Corpses," 28.

Bibliography

Allen, Anita. "Body Ethics, Body Aesthetics." *Philadelphia Inquirer*, March 12, 2006. "Body Language." Philoillogica. http://philoillogica.typepad.com/philoillogica/2006/01/body_language.html

Institute for Plastination. Gunther von Hagens' Body Worlds: The Anatomical Exhibition of Real Human Bodies. Heidelberg: 2006. http://www.bodyworlds.com

_____."Questions & Answers." Institute for Plastination. Gunther von Hagens'. http://www.bodyworlds.com/en/exhibitions/questions_answers.html

_____."The Plastination Process." Institute for Plastination. Gunther von Hagens'. http://www.bodyworlds.com/en/plastination/idea_plastination.html

Cho, Margaret. "Body Worlds." Margaret Cho's Blog. http://www.margaretcho.com/blog/bodyworlds.htm

Claire, Christopher. "Paint Protest Over Corpses Exhibition." *The Scotsman*, March 24, 2002. http://www.western.edu (accessed Jan. 12, 2007).

Hall, Stuart. "Ethnicity: Identity and Difference." In *Border Texts*, edited by Randall Bass, 295–305. New York: Houghton, 1999.

Haraway, Donna. *Simians, Cyborgs, and Women: The Reinvention of Nature.* New York: Routledge, 1991.

Hirschauer, Stefan. "Animated Corpses: Communicating with Post Mortals in an Anatomical Exhibition." *Body & Society* 12, no. 4 (2006): 25–52.

Jally, Sut. *Stuart Hall: Race: The Floating Signifier*. Massachusetts: Media Education Foundation, 1997.

Johnson, Stephen. "My Body in Plastic: A Tourist Goes to Body Worlds." *Canadian Theatre Review* 127 (2006): 74–80.

Linke, Uli. "Touching the Corpse: The Unmaking of Memory in the Body Museum." *Anthropology Today* 21, no. 5 (2005): 13–19.

Moore, Lisa Jean, and Adele E. Clarke. "Clitoral Conventions and Transgressions: Graphic Representations in Anatomy Texts, c1900–1991." *Feminist Studies* 2, no. 2 (1995): 255.

Rose, Steven. *Lifelines: Biology Beyond Determinism*. New York: Oxford University Press, 1998.

Stern, Megan. "Dystopian Anxieties Versus Utopian Ideals: Medicine from *Frankenstein* to *The Visible Human Project* and *Body Worlds*." *Science as Culture* 15, no. 1 (2006): 61–84.
von Hagens, Gunther. "Anatomy and Plastination." In Whalley, *Gunther von Hagens'*, 9–37.
_____. "On Gruesome Corpses, Gestalt Plastinates and Mandatory Interment." In Whalley, *Gunther von Hagens'*, 260–82.
_____. "Why I Think It Is High Time We Democratised the Science of Anatomy." *Times Higher Education Supplement*, March 22, 2002. http://www.timeshighereducation.co.uk/story.asp?storyCode=167963&sectioncode=26 (accessed January 15, 2007).
Walter, Tony. "Body Worlds: Clinical Detachment and Anatomical Awe." *Sociology of Health & Illness* 26, no. 4 (2004): 464–88.
_____. "Plastination for Display: A New Way to Dispose of the Dead." *Journal of the Royal Anthropological Institute* 10, no. 3 (2004): 603–27.
Whalley, Angela, ed. *Gunther von Hagens' Body Worlds: The Anatomical Exhibition of Real Human Bodies*. Trans. Francis Kelly. Heidelberg: Arts & Sciences, 2006.

The Politics of Fetal Display

Christian DuComb

Death is the release from all pain and complete cessation, beyond which our suffering will not extend. It will return us to that condition of tranquility which we had enjoyed before we were born. Should anyone mourn the deceased, then he must also mourn the unborn.

— Seneca

The passage above, attributed to Seneca, hangs on a large banner over the first gallery of Body Worlds 2, which my wife Caitlin and I visited at the Museum of Science, Boston, in November 2006. Subtitled "the anatomical exhibition of real human bodies" and advertised on billboards throughout the Boston area as "original," "authentic," and "inspiring," Body Worlds 2 displays cadavers preserved through plastination, a "patented method of anatomical preservation [that] replaces bodily fluids and fat in donor specimens with reactive fluid plastics."[1]

Hailed in the exhibition brochure as a "pioneering technology," plastination — developed by German anatomist and impresario Gunther von Hagens—provides the material support for most but not all of the exhibits in Body Worlds 2. These exhibits can be divided into roughly four categories: 1. plastinated organs and sagittal cross-sections of human cadavers, organized in display cases to demonstrate the workings of the locomotive, nervous, respiratory, cardiovascular, digestive, and reproductive systems; 2. whole body "plastinates" fixed in what the exhibition brochure describes as "dynamic, lifelike poses" and signed by von Hagens as works of art; 3. animal displays, including a plastinated camel; and 4. human embryos and fetuses from historical collections, preserved using methods that predate plastination.

I will concentrate my analysis on the last of these four categories, but to do so, I must situate the practices of fetal display in Body Worlds 2 in both the specific context of von Hagens' exhibition and the larger context of what Meredith W. Michaels and Lynn M. Morgan have called "the burgeoning public fascination with fetuses"[2]—a fascination that both emerges from and conditions political debates about abortion in the United States.*

Through a comparison of the fetal exhibits in Body Worlds 2 and the whole body plastinates that dominate the exhibition, I argue that Body Worlds 2 reinforces the agenda of the anti-abortion movement by encouraging visitors to *identify* with the fetuses on display. The Reproductive Development Gallery near the end of Body Worlds 2, which exhibits fetuses and embryos from historical collections, radically reverses the sense of distance the viewer feels from the whole body plastinates in the rest of the exhibition. For Petra Kuppers, who critiques the original Body Worlds from a disability studies perspective, this sense of distance results from von Hagens' erasure of individual particularity from the bodies he preserves through plastination. I combine this argument with Elizabeth Grosz's theorization of the freak to suggest that Body Worlds 2 participates in the representational economy of the freak show, further increasing the viewer's sense of distance from the bodies on display. By contrast, the Reproductive Development Gallery encourages not distance from its exhibits but identification with them; and this identification contributes to what Sarah Franklin has called "fetal personhood," the misrecognition of the fetus as an autonomous form of life. In Franklin's view, the illusion of fetal personhood rests on the composition and framing of *images* of the fetus. The final section of this essay interrogates the distinction between fetal *images* and fetal *bodies*, ultimately concluding that the latter—when displayed ex utero as in Body Worlds 2—present the viewer with an intractable ambiguity between the dead and the unborn. This ambiguity makes the Reproductive Development Gallery, more so than any other aspect of Body Worlds 2, a freak show *par excellence*, with more complex political implications than an analysis limited to fetal personhood would suggest.

I. Body Worlds 2 as Freak Show

Together, the original Body Worlds and its sequel exhibitions Body Worlds 2 and Body Worlds 3 have attracted nearly twenty-five million visitors in Europe, Asia, and North America.[3] But despite its popularity, scholarly responses to Body Worlds have tended to be negative. In her salient critique, Kuppers takes the original Body Worlds to task for submitting the bodies on display to conventions of representation that erase the individual, lived experience of the subjects who once inhabited them. The "lifelike poses" of the exhibition's whole

**I first undertook to write about the fetal exhibits in Body Worlds 2 in Karen Shimakawa's seminar "Freakshows: Racialized Performance" at Brown University in 2006, and I benefited greatly from many of Karen's suggestions through several drafts of this essay. In addition, I presented parts of this essay at the Performance Studies International conference at New York University in 2007 and learned much from the feedback I received there.*

body plastinates conform closely to gender stereotypes—male athletes and female dancers abound—and von Hagens flays the skin of these exhibits to reveal their internal anatomy—a choice that strips away each body's racial, cultural, and historical specificity. As a result, Body Worlds shows the human body on what Kuppers calls "a strange edge: the moment when the index of the body covers its living space."[4] This bloated indexicality, which Kuppers likens to a map covering the whole of the territory it describes, turns Body Worlds into "a realist stage where corpses [stand] for people."[5] Here, Kuppers cites Elin Diamond's feminist critique of stage realism as a style that mystifies the process of theatrical signification:

> Because it naturalizes the relation between character and actor, setting and world, realism operates in concert with ideology. And because it depends on, insists on a stability of reference, an objective world that is the source and guarantor of knowledge, realism surreptitiously reinforces (even if it argues with) the arrangements of that world.[6]

In other words, realism elides the distinction between reality and representation, and the hyper-realism of the "real human bodies" on display in Body Worlds occludes the fact that these bodies *stand* for people.

Building on Kuppers' argument, I would submit that Body Worlds 2 functions as a "realist stage" that mystifies the process of theatrical signification while also pointing to and celebrating its own artifice. Publicity materials highlight plastination as the enabling condition of the exhibition, "the pioneering technology that makes contemporary anatomical exhibits possible"; and the juxtaposition of whole body plastinates with reproductions of Renaissance anatomical drawings locates the plastinated displays within a specific tradition of representing the human body. In this sense, Body Worlds 2 touches both sides of the paradoxical notion of theatricality: it pretends to transparency, to showing things as they "really are," while simultaneously flaunting the material support and aesthetic conventions that constitute it. The doubleness of the exhibition's theatricality also doubles the viewer's sense of distance from the bodies on display. The plastinated cadavers masquerade as "real," repressing the historical particularity of the people they once were, the "stories of life and flesh" that Kuppers longs to hear;[7] and at the same time, these exhibits also direct the viewer's attention to the *staging* of the bodies on display and away from the bodies themselves.

On a similar banner to the one that greets visitors to Body Worlds 2 with a passage from Seneca, von Hagens quotes himself: "Plastination unveils the beauty beneath the skin, frozen in time between death and decay." In Kuppers' view, plastination serves not as a technology of unveiling but as a realist representational technique that produces "corpse simulations."[8] And these simulations, she argues, "partake in a strange dance of death ... [that] attempts to disguise death's temporality, the fact that life *and* death are processes that depend on change and transformation."[9] Plastination arrests the body in its passage across the limit between life and death, freezing it in time, drying it out, flaying its skin to reveal its insides by making the person who *was* that body unrecognizable. For von Hagens, this preservation process "transforms the body, an object of individual mourning, into an object of reverence, learning, enlightenment, and appreciation."[10] But in so doing, it denies death as a messy, physical, social, and *temporal* process, instead rendering it neat and clinical, a fixed and frozen moment for public exhibition where the true "object of reverence" is not the body itself but the technology, the theatrical artifice, that preserves it.

Body Worlds 2 also suspends the body at the threshold between life and death by staging its whole body exhibits in arrested motion — in the midst of kicking a soccer ball or dancing *en pointe*, to cite two examples. And through this second form of temporal arrest, I would argue that the exhibition participates in the representational economy of the freak show. Peeling away the skin, the heaviest human organ, to expose the ordinarily hidden structures that lie beneath it not only excites the viewer's voyeuristic gaze but also "confirms the viewer as bounded, belonging to a 'proper' social category"[11] — the two audience responses that Grosz situates as constitutive of freak show performances and displays. "The freak," writes Grosz,

> is an *ambiguous* being whose existence imperils categories and oppositions dominant in social life. Freaks are those human beings who exist outside and in defiance of the structure of binary oppositions that govern our basic concepts and modes of self-definition. They occupy the impossible middle ground between the oppositions dividing the human from the animal (Jo-Jo, the dog-faced boy; Percilla, the monkey girl; Emmitt, the alligator-skinned boy; the "wild man" or "geek"), one being from another (conjoined twins, "double-bodied wonders," two-headed or multiple-limbed beings), nature from culture (feral children, the "wild men of Borneo"), one sex from the other (the bearded lady, hermaphrodites, Joseph-Josephines or Victor-Victorias), adults and children (dwarfs and midgets) humans and gods (giants) and the living and the dead (human skeletons).... They imperil the very definitions we rely on to classify humans, identities, and sexes — our most fundamental categories of self-definition and boundaries dividing self from otherness.[12]

By presenting dead bodies frozen in lifelike poses, Body Worlds 2 places its whole body exhibits in suspended *re*-animation, creating an ambiguity between the living and the dead. This ambiguity ultimately serves to confirm the viewer as a bounded, living self by depicting death as "other" and frozen in time — or, more precisely, *outside* of time. And this strange temporal arrest, compounded with the exhibition's double theatricality, only serves to deepen the viewer's sense of distance from the bodies on display.

Of course, the freak show does more than confirm the viewer as "bounded"; it also excites a flash of horror, which Grosz locates in the viewer's recognition that the freak is "at the heart of his or her own identity, for it is all that must be ejected or abjected from self-image to make the bounded, category-obeying self possible."[13] By placing death outside of time, Body Worlds 2 almost succeeds in repressing the social, temporal, and even physical realities of death. But the undeniable fact that the bodies on display *were once alive* subsists in the exhibition as a *memento mori*, inciting a tiny panic that impels the viewer to abject the freakish, plastinated exhibits from her sense of self. The specially marked gallery of fetuses and embryos near the end of the exhibition radically reverses the viewer's by-now reflexive sense of distance from the bodies on display. Instead, Body Worlds 2 brings us *close* to the fetus and embryo exhibits in the Reproductive Development Gallery, positing conception as the limit where life begins just as certainly as it posits death as a clear limit outside the physical, social, and temporal processes that constitute it. In its presentation of embryos as "young" as four weeks, Body Worlds 2 undercuts one side of Seneca's wisdom: don't mourn the dead, it tells us; mourn the unborn.

II. "Fetal Personhood" and Abortion Politics

In 1997, science historian and cyber-feminist Donna Haraway remarked that "it is almost impossible to get through the day near the end of the Second Christian Millennium in the United States without being in communication with the public fetus."[14] She wasn't exaggerating. As Valerie Hartouni observes, fetal ultrasound images appeared in advertisements for "cars, bottled water, sitcoms, and long distance telephone services" over the course of the 1990s, while "the circulation of [Swedish photographer Lennart] Nilsson's arresting, prenatal portraits" fueled the sometimes violent protests of the anti-abortion movement.[15] Franklin points out that many of the fetuses pictured in Nilsson's photographs—which first appeared in *LIFE* magazine in 1965, and which have circulated widely since—were recovered from late-term abortions, making their appropriation by anti-abortion campaigns ironic and arguably hypocritical.[16] Franklin maintains that "these portraits of fetal tranquility (actually portraits of fetal post-mortems) ... offer a specific set of points of identification with the fetus," enhanced by the captions that usually accompany the images in anti-abortion propaganda.[17] In other words, the textual framing of Nilsson's photographs in many of the anti-abortion tracts where they appear simultaneously constructs and conceals a freakish ambiguity between the living and the dead evident in the tension between the purported referent—tranquil fetus, in utero—and the actual referent—dead fetus, ex utero. This sleight of hand transforms Nilsson's hyper-real images of the fetus into sites of identification, into pictures in which we recognize ourselves.

Franklin contends that a sense of *fetal personhood* emerges from this identification, underwritten by two distinct processes of separation: "[the] separation ... of the fetus from the mother, and of the social from the biological."[18] The first of these two separations inheres in the composition of the images themselves, where

> the position of the spectator ... is most significantly defined by the structuring absence of the mother, who is replaced by empty space.... Set against a black backdrop, the round fetal orbs [in Nilsson's photographs] resemble nothing more than a sort of organic spacecraft, floating in a void. Through this technique, the conditions for spectator recognition of and identification with the fetus are achieved on the basis of its separation and individuation from the mother.[19]

Although Scott Gilbert correctly points out that Nilsson's book, *A Child is Born*, "is adamant in portraying the embryo and fetus as integrated into the woman's body ... and woman's life,"[20] Nilsson's photographs nonetheless show the fetus in isolation, the sole criterion on which the rhetorical use of his images by the anti-abortion movement depends. For instance, an image of an eighteen-week fetus appears in his book opposite an image of a woman in the eighteenth week of pregnancy; but the former image is reproduced out of context on the anti-abortion website Actionlife.org — without the accompanying photo of the pregnant woman — "to show the innocence and purity and childlike nature of the pre-born baby."[21]

The style of presentation of most, though not all, of the fetuses displayed in Body Worlds 2 echoes the latent tendency of Nilsson's photographs to isolate the fetus from the mother. A sign in the penultimate gallery of the exhibition invites visitors to step into a curtained room and "experience the wonders of human development." Across the threshold, the lights dim and ambient music plays; thick red drapes hang along the walls, creating a sense of enclosure and intimacy markedly different from the open, airy arrangement of all the preceding galleries. Figuratively speaking, it feels as though one has stepped (back) into the womb. This womb/room exhibits four fetuses ex utero, at the nineteenth/twentieth, twenty-second, twenty-fourth, and twenty-eighth weeks of development. Each fetus lies in its own display case, cushioned by thick black fabric and lit in such a way that it almost seems to be glowing. Adjacent to these fetal displays are nine embryos ranging from four to eight weeks of development, each suspended in a thin, luminous tube.

A placard at the entrance to the Reproductive Development Gallery explains that "to the best of our knowledge, the human embryos and fetuses displayed here (from historical collections over eighty years old) failed to survive to birth because of natural accidents." Although the provenance of the four fetuses in question may make the display of their internal organs and musculature impossible, the fact that they are the *only* intact, whole bodies presented in the entire exhibition compels the viewer to identify with them. As evinced elsewhere in Body Worlds 2, technological limitations do not oblige von Hagens to exhibit intact fetuses from historical collections rather than newly plastinated fetal bodies with their internal anatomy exposed. Just beyond the curtained womb/room of the Reproductive Development Gallery, two plastinated fetuses appear in a display case with almost no commentary to contextualize them. In one of these fetal exhibits, a translucent specimen in the fifth month of development, "the bones appear through the soft tissue due to special coloration during plastination."[22] Like the whole body adult plastinates elsewhere in the exhibition, its insides are visible, situating it in the same freak show economy that distances the viewer from most of the displays in Body Worlds 2. The second plastinated fetus—shown in utero in the third month of pregnancy with its mother's body nowhere in sight—seems to belong in the womb/room next door, where it would fill the gap in the specimens displayed there between the eighth and nineteenth weeks of embryonic and fetal development. But I would contend that a fetus at three months, which looks more like a prawn than a baby, might disrupt the viewer's identification with the four intact, glowing, and well-developed fetal bodies carefully displayed in the Reproductive Development Gallery.

By representing the fetus as a discrete entity, both Nilsson's photographs and the fetal exhibits in the Reproductive Development Gallery obscure the material relationship between the *live* fetus and the mother; and when represented outside the uterus that once sustained it, the *dead* fetus seems, strangely, to take on a life—or at least an existence—of its own. The fetus in these instances emerges as an "individual," a word for which Franklin reserves special vitriol in her critique of fetal personhood:

> The very term "individual," meaning *one who cannot be divided*, can only represent the male, as it is precisely the process of one individual becoming two which occurs through a woman's pregnancy. Pregnancy is precisely about one body becoming two, two bodies becoming one, the exact antithesis of individuality.[23]

In other words, pregnancy—through the lens of my expansive reading of Grosz's theorization of the freak—is a kind of freak show, an ambiguous embodiment that troubles the binary opposition of "one being from another." But unlike most other forms of "freakery"—conjoined twins, for example—pregnancy is an inherently transient phenomenon, and the fact that it will *end* mitigates the threat it poses to the concept of individual, bounded subjectivity.

Nonetheless, pregnancy raises the specter of one-in-two and two-in-one, a freakish ambiguity that displays of fetal personhood smooth over when projected back onto the pregnant body. The image of the fetus as a separate individual translates pregnancy into a condition that fits more comfortably within what Grosz calls "the structure of binary oppositions that govern our basic concepts and modes of self-definition." But by separating the fetus from the mother, images of fetal personhood arrest the temporal flow of pregnancy, isolating the fetus outside of time much as the adult whole body plastinates in Body Worlds 2 strip death of its temporality. The ontology suggested by this temporal arrest gives the fetus—and, by extension, the embryo—an *a priori* "right to life" *independent* of its mother. From this position, the false principle that life begins at the moment of conception emerges as a logical necessity. Beyond its dangerous implications for the abortion debate, this line of reasoning places women under erasure by emphasizing one party in pregnancy, the fetus, over the other. As Franklin puts it, "endowing fetuses with full civil rights ironically confers upon them a status ... *which women never had to begin with.*"[24] And as importantly, the ontology suggested by images of fetal personhood overlooks embryonic and fetal development as highly mediated processes in which the limit where "life" begins shifts with the changing tides of medical research, bioethical and religious debate, and legal wrangling over the regulation of stem cell lines and access to abortion.

Fetal personhood is an illusion predicated on images of dead fetuses, and it seems obvious that showing the fetus *with* the mother would upend conventions of fetal representation and reclaim what Franklin calls a woman-centered perspective on reproduction as "a social process involving social persons, who are interdependent and whose right to exist is not bound up with notions of radical separateness."[25] Before it presents four glowing, individuated fetuses, each in its separate display case, Body Worlds 2 confronts the viewer with an apparent *détournement*: a plastinated woman shown with her plastinated fetus in utero. This exhibit seems to resolve the separation of fetus and mother endemic to images of fetal personhood through its presentation of a mother-fetus pair. The woman-cum-fetus exhibit ambiguously straddles two of the binary oppositions cited by Grosz as part of the dominant social structure challenged by the freak: the opposition of "the living and the dead," and the opposition of "one being from another." In the first of these binaries, the woman-cum-fetus exhibit operates much like the other plastinates in Body Worlds 2: its life/death ambiguity serves as an easy foil for the viewer to consolidate her bounded sense of self. However, in the opposition of one being from another, the woman-cum-fetus seems to present not only a more stubborn ambiguity but also a potential site of political resistance, where the twoness of pregnant embodiment might overcome the binary that would split it. But a third opposition, one not theorized by Grosz, cuts against the resistant potential of the exhibit: the binary of *outside* and *inside.*

The flayed skin of the plastinated woman reveals the inside of her body, where an intact five-month old fetus, most of its developing dermis still in place, rests curled up in utero. This intact body-within-a-body tips the scales of the opposition between one being and another; rather than two interdependent bodies in one, we see a discrete fetal individual which looks almost ready to spring from within its dissected mother. An image of fetal personhood rises out of the exhibit, resolving the ambiguity of two-in-one into clear roles of dominance and submission and turning the opposition of outside and inside inside out. A review by Jason Byassee in the conservative Christian press of the first Body Worlds, which also contains a woman-cum-fetus plastinate, captures the sinister politics of this particular practice of fetal display: "I found myself drawn by this ... exhibit. I not only felt for the child who would never be born, but felt claustrophobic at seeing the fetus scrunched inside his plastinate mother.

The exhibit's intention of blurring the line between "inner" and "outer" is achieved to a breathtaking degree."[26] Byassee's identification with and *mourning* over the fetus—which he calls a "child"—coupled with his relative indifference to the mother—whom he calls a "plastinate"—makes both this exhibit's complicity in the illusion of fetal personhood and its implications for abortion politics abundantly clear.

On our way into the Reproductive Development Gallery, Caitlin and I brushed past a middle-aged woman and her young daughter, perhaps ten years old, exiting the curtained room that we were about to enter. I overheard a snippet of their conversation: "People argue over when life begins," said mother to daughter; "it's hard to argue after seeing something like that." ("Like *what*?" I wondered). Tugging on Caitlin's sleeve to find out if she heard what I did, we found ourselves suddenly in front of the mother-cum-fetus exhibit, the first thing the visitor sees upon walking into this gallery. The display captivated me: before the opaque flesh of a pregnant body rendered transparent, I felt I was peering into a great mystery. For the first time all day, I saw something in one of the exhibits, behind the skin, with which I could *identify*—a feeling deepened, no doubt, by the fact that Caitlin was eight weeks pregnant at the time. We moved along, following the foot traffic counterclockwise through this womb/room whose dim intimacy and soothing music hushed the crowd and provided a welcome respite from the rest of the exhibition. As we traveled a reverse course through the process of fetal development—from the fetus in utero in the woman-cum-fetus plastinate, to the four individual "fetal persons," to a series of tadpole-sized embryos—the feeling of identification with the fetus traveled with me. The tension between my professed politics and my affective response to these exhibits troubled me deeply and raised a question that has lingered ever since: did the prospect of fatherhood awaken powerful hormones and deep-seated acculturations that caused me to identify with the fetuses on display, or did the politics of the exhibition insinuate themselves into my response? Although I do not discount the former explanation, the comment I overheard as Caitlin and I entered the womb/room suggests that the Reproductive Development Gallery has exercised a coercive and politically dangerous influence on at least some of the millions of people who have seen it.

III. "Reproductive Futurism" and the Fetal Body

The elision of the distinction between fetus and child appears not only in Byassee's response to Body Worlds but also in the curatorial language of the exhibition itself. A wall-mounted placard on embryonic and fetal development in the Reproductive Development Gallery in Body Worlds 2 proclaims that "by the end of the eighth week, all of the organs will be in place ... after which the developing child is referred to as a fetus." The gratuitous and inaccurate use of the word "child" in reference to the medical distinction between a fetus and an embryo renders explicit the participation of images of fetal personhood in what queer theorist Lee Edelman calls "reproductive futurism," the all-pervasive ideology of the Child outside of which he maintains political discourse cannot exist:

> Historically constructed ... to serve as the repository of variously sentimentalized cultural identifications, the Child has come to *embody* for us the telos of the social order and come to be seen as the one for whom that order is held in perpetual trust. In its coercive universalization, however, the *image* of the Child, not to be confused with the lived experiences of any historical children, serves to regulate political discourse — to prescribe what will *count* as political discourse — by compelling such discourse to accede in advance to the reality of a collective future whose figurative status we are never permitted to acknowledge or address.[27]

Edelman's critique of the Child as futurity — encapsulated in the passage above from his essay "The Future is Kid Stuff" — slips from discussing what the Child *embodies* to treating the Child as an *image*, a problem I take up below. But this slippage does not detract from his larger points that all political discourse must take place within the ideology of the Child, and that "*queerness* names ... the side outside the consensus by which all politics confirms the absolute value of reproductive futurism."[28] Rather than join this consensus, Edelman stakes a claim to a radical queer position which insists that "the Child as futurity's emblem must die; [and] that the future is mere repetition, and just as lethal as the past."[29] His assertion of a queer subjectivity outside the social order troubles the feminist view — articulated by Franklin among others — that personhood is, or at least should be, constructed in the realm of the social.

Indeed, Edelman hardly counts feminists as allies in his project:

> Even proponents of abortion rights, while promoting the freedom of women to control their own bodies through reproductive choice, recurrently frame their political struggle, mirroring their anti-abortion foes, as a "fight for our children — for our daughters and our sons," and thus a fight for the future.[30]

This criticism correctly diagnoses the effort of the "pro-choice" movement, embedded in the name by which it calls itself, to displace attention from the *practice* of abortion to the more politically palatable issues of abortion access and abortion rights. To be *pro*-abortion would be to *oppose* the future and to acknowledge without flinching the dead, aborted fetuses that the pro-choice movement would rather hide; in Edelman's terms, it would amount to a renunciation of politics in favor of a queer negativity. In a response to John Brenkman's critique of "The Future is Kid Stuff," Edelman makes the affinity between queerness and abortion explicit. "The force of negativity," he writes, "associated in our moment with queerness and abortion both, finds few who will espouse it; a universal futurism precludes our unambiguously taking its side. But the need to *embody* that force in some *figure* of anti-sociality remains" (my emphasis).[31] Here again, Edelman slips from the language of embodiment into the language of figure and image, leaving the reader to wonder what actual bodies (if any) he means to bring into play in his discussion. Queer bodies, certainly — by which he means "[anybody] so stigmatized for noncompliance with heteronormative mandates"[32]–but I would submit that the aborted fetus embodies "the force of negativity" as much as, though differently than, the queer.

However, the intervention of the fetal body at this juncture in Edelman's argument threatens to shift fetal personhood —constructed through images of dead fetuses—from the side of reproductive futurism, where it obviously belongs, to the side of queerness. The absurdity of this threat brings into focus the distinction between fetal *images* and fetal *bodies*, a distinction essential to understanding the politics of fetal display in Body Worlds 2. As Franklin shows, the construction of fetal personhood operates in similar ways wherever "the public fetus" appears. But the fetal bodies in the Reproductive Development Gallery— unlike the fetal images in Nilsson's photographs—cannot be completely subsumed by an ideology, neither the ideology of fetal personhood nor the ideology of the Child. Although the fetuses on display in Body Worlds 2 certainly participate in these ideologies, their third dimension, their *dead flesh*, points both to the arrested temporal flow of their development and the impossibility of its realization. This thwarted potential suggests a future that might have done more than merely repeat the past and which, for precisely this reason, can never be known. The latent, failed promise that clings to the fetal bodies on display in Body Worlds 2 "queers" reproductive futurism by positing the impossible possibility of a future *with a difference*. It also vexes abortion politics, forcing even feminists like Morgan and Michaels (whom I cited at the beginning of this essay) to acknowledge that the fetus merits a place on the social stage.[33] The deep ambiguities of the fetal body — between repetition and difference, the dead and the unborn, the promise of the future and the future's impossibility — make the exhibition of the fetus the ultimate freak show. But divested of the technologies and practices that make it visible and seem to make it speak, the fetus is invisible and silent: it is neither subject nor object, its experience remains unknowable, and its multiple ambiguities dissolve all oppositional structures of thought and refuse to return us to a neat, bounded sense of our selves.

Notes

1. Museum of Science, Boston, *Body Worlds 2*.
2. Morgan and Michaels, "Introduction: The Fetal Imperative," 1.
3. Institute for Plastination, "Body Worlds: An Anatomical Success Story."
4. Kuppers, "Visions of Anatomy," 128.
5. Ibid., 151.
6. Diamond, *Unmaking Mimesis*, 4–5.
7. Kuppers, 140.

8. Ibid.
9. Ibid.
10. Gunther von Hagens', Exhibition Placard.
11. Grosz, "Intolerable Ambiguity: Freaks as/at the Limit," 65.
12. Ibid., 57.
13. Ibid., 65.
14. Cited in Hartouni, "Epilogue: Reflections on Abortion Politics and the Practices Called Person," 302.
15. Ibid.
16. Franklin, "Fetal Fascinations: New Dimension to the Medical-Scientific Construction of Fetal Personhood," 195.
17. Ibid.
18. Ibid., 196.
19. Ibid., 195–6.
20. Gilbert, "Images of Embryos Used by Anti-Abortion Activists."
21. Ibid.
22. Gunther von Hagens'.
23. Franklin, 203.
24. Ibid., 201.
25. Ibid., 203.
26. Byassee, "The Human Specimen," 10.
27. Edelman, "The Future Is Kid Stuff," 10–11 (emphasis added).
28. Ibid., 3.
29. Ibid., 31.
30. Ibid., 3. Edelman quotes Donna Shalala, who served as Secretary of Health and Human Services under Bill Clinton.
31. Edelman, "Post-Partum," 185 (emphasis added).
32. Ibid., 182–3.
33. Morgan and Michaels, 5.

Bibliography

Byassee, Jason. "The Human Specimen." *Christian Century*, March 22, 2005.
Diamond, Elin. *Unmaking Mimesis*. London: Routledge, 1997.
Edelman, Lee. "The Future Is Kid Stuff." In *No Future: Queer Theory and the Death Drive*, 1–32. Durham: Duke University Press, 2004.
_____. "Post-Partum." *Narrative* 10, no. 2 (2002): 181–185.
Franklin, Sarah. "Fetal Fascinations: New Dimension to the Medical-Scientific Construction of Fetal Personhood." In *Off-Centre: Feminism and Cultural Studies*, edited by Sarah Franklin, Celia Lury, and Jackie Stacey, 190–205. London: Harper Collins Academic, 1991.
Gilbert, Scott F. "Images of Embryos Used by Anti-Abortion Activists." DevBio: A Companion to Developmental Biology. http://8e.devbio.com/article.php?id=164 (accessed January 22, 2008).
Grosz, Elizabeth. "Intolerable Ambiguity: Freaks as/at the Limit." In *Freakery: Cultural Spectacles of the Extraordinary Body*, edited by Rosemarie Garland Thomson, 55–66. New York: New York University Press, 1996.
Gunther von Hagens' Body Worlds 2: The Original Exhibition of Real Human Bodies. Exhibition Placard. Museum of Science, Boston.
Haraway, Donna J. *Modest_Witness@Second_Millennium. FemaleMan_Meets©_Oncomouse™*. New York: Routledge, 1997, cited in Hartouni.
Hartouni, Valerie. "Epilogue: Reflections on Abortion Politics and the Practices Called Person." In *Fetal Subjects, Feminist Positions*, edited by Lynn M. Morgan and Meredith W. Michaels, 296–303.
Institute for Plastination. "Body Worlds: An Anatomical Success Story." Gunther von Hagens'

Body Worlds: the Original Exhibition of Real Human Bodies. Heidelberg: 2007. http://www.bodyworlds.com/en/prelude.html (accessed September 26, 2007).

Kuppers, Petra. "Visions of Anatomy: Exhibitions and Dense Bodies." *differences: a journal of feminist cultural studies* 15, no. 3 (2004): 123–156.

Morgan, Lynn M., and Meredith W. Michaels. "Introduction: The Fetal Imperative." In *Fetal Subjects, Feminist Positions*, edited by Lynn M. Morgan and Meredith W. Michaels, 1–9. Philadelphia: University of Pennsylvania Press, 1999.

Museum of Science, Boston. Gunther von Hagens' Body Worlds 2: The Original Exhibition of Real Human Bodies. Exhibition Brochure. Boston: 2007.

Corpse-less: A Battle with Abjection

Elizabeth Simon Ruchti

Introduction

Is it possible to make a room full of artistically displayed human corpses, internal organs, muscle tissue, and sheets of skin anything but revolting to normative sensibilities? Is there a way to make death safe and corpses entertaining? Apparently yes. Millions of people around the world have clamored to see Gunther von Hagens' displays of anatomical specimens in Body Worlds. Using his preservation technique, known as *plastination*, von Hagens transforms human corpses into plastic-filled, life-sized action figures. By replacing body fluids with liquefied plastic, plastination halts the body's natural process of decay and transforms the corpse into a plastinate — a safe, sterile version of a corpse, free of the stench of rot and any reminders of our mortality.[1] No longer corpses, the anatomical specimens of Body Worlds take on a theme-park-like aesthetic at times vaudevillian and, at others, freak-show-esque. In other words, von Hagens has stripped the corpse of all of its abject *corpse-ness*. That is, he removes the real yet abject and replaces it with the ideological yet tangible version of our immortality.

The plastination technique, and Body Worlds' staging of the plastinated specimen, neutralizes the most disturbing elements of the corpse. The plastinated corpse appears somewhat human, human enough that audiences can relate to, identify with, and — most importantly — not fear it the way that we fear a corpse. At the same time, we must not identify with the plastinate too much, not so much that we lose our objectivity, our ability to distance ourselves from the plastinated object before us. We cannot relate to it so much that we begin to fear it as part of ourselves. For what is it that we fear in the corpse? We fear the knowledge that inside each of us is a rotting, stinking corpse waiting to get out. We fear the corpse because the corpse is not at all like us; and we fear the corpse because the corpse *is* us. The corpse is simultaneously object — not like us — and subject — us. It is the ultimate *abject* body. The plastinate, however, is not an abject body. It is not a corpse, because it is *like* us, but is *not us*.

In this essay, I argue that Gunther von Hagens normalizes the corpse, neutralizes its abject power, its queerness, by plastinating and staging it as the normative subject's *other*. Plastination stabilizes the process of decay inherent to the corpse, thereby allowing von Hagens to control the meanings generated by the once unwieldy corpse. In other words, he de-queers the abject: he takes what is most abject to normative society and attaches normative identity markers to it. His most obvious attempts to define and thereby de-queer the plastinate can be found in the abundance of gender norms imposed on the displays. Von Hagens attaches heteronormative tropes to the cadaver to rid it of its abjectifying power over the audience. I contend, however, that while von Hagens succeeds in de-abjectifying, or, in other words, *othering* the corpse, he does so only because the plastinated corpse is no longer a corpse. Von Hagens fails in his quest to nullify the abject power of the *actual* corpse because the object he presents in his exhibits is no longer a corpse but a safely contained object called a plastinate.

Through the work of Julia Kristeva, I explain the power, danger, and beauty of abjection. Queering the subject-other dichotomy through the abject is powerful in part because of the lengths to which communities and individuals go to rid themselves of any connection to the abject. Abjection is dangerous because no matter how hard communities and individuals try, they will never escape its threat. Most importantly, however, the abject is beautiful because for all the potential harm communities and individuals see in un-harnessed abjection, it is abjection's un-harness-ability that ensures them infinite options for identification and self-definition. This becomes most obvious when considered in relation to Gunther von Hagens' Body Worlds exhibits and his conscious efforts to normalize — and in effect de-abjectify — the corpse. The normative tropes von Hagens employs in his staging of Body Worlds demonstrate the resilience of the abject, his impotent attempts to control and squelch the power of the abject, and the abject's inevitable specter-like presence in subject formations.

The Ever-Allusive Abject

In *Powers of Horror: An Essay on Abjection*, Julia Kristeva begins her explanation of the abject by stating what it is not: "The abject is not an ob-ject facing me, which I name or imagine. Nor is it an ob-jest, an otherness ceaselessly fleeing in a systematic quest of desire. What is abject is not my correlative, which, providing me with someone or something else as support, would allow me to be more or less detached and autonomous."[2] Kristeva introduces the abject by way of negatives because the abject *is not*. The abject is *not* the subject, but it is *in* the subject. It is a thing, but it is not definable as *some*thing, as an object that opposes the subject. Yet, the abject does oppose the subject, but not as an object, not as a definable something. These contradictions are central to Kristeva's explanation of the abject:

> The abject has only one quality of the object — that of being opposed to *I*. If the object, however, through its opposition, settles me within the fragile texture of a desire for meaning, which as a matter of fact, makes me ceaselessly and infinitely homologous to it, what is *abject*, on the contrary, the jettisoned object, is radically excluded and draws me toward the place where meaning collapses. A certain ego that merged with its master, a superego, has flatly driven it away. It lies outside, beyond the set, and does not seem to agree to the latter's rules of the game. And yet, from its place of banishment, the abject does not cease challenging its master.[3]

The object exists as the *other* of the subject, opposed to I. We define ourselves in opposition to our other. We seek knowledge through our understanding of the difference between our subjectivity and the objects' otherness. The abject, however, does not participate in this dichotomy. The abject's opposition to the subject threatens the stability of the subject's identity, for the abject is what the subject *is*, but what the subject can *never be*.

Excrement, refuse, and corpses serve as Kristeva's primary examples of the abject: "...refuse and corpses show me what I permanently thrust aside in order to live."[4] Each of us, everyday, produces waste and becomes waste. We are waste. But we must also reject waste, avoid it, and deny its constant presence to affirm our desire to continue living, growing, not-decaying. For Kristeva, the balance the subject strikes between growth and decay, life and death, exemplifies the abject: "There I am at the border of my condition as a living being. My body extricates itself, as being alive, from that border. Such wastes drop so that I might live, until, from loss to loss, nothing remains in me and my entire body falls beyond the limit—*cadere*, cadaver. If dung signifies the other side of the border, the place where I am not and which permits me to be, the corpse, the most sickening of wastes, is a border that has encroached upon everything."[5]

We have a profound fear of the corpse because we see our inevitable future in its rot. We also recognize our present, a present we hide from always. Each day our bodies decay. We slough off this decay — sweat, feces, skin — so it cannot contaminate what remains healthy in us. We hide our decay and hide from it. We avoid letting others come into contact with our decay. We avoid their decay and our own. Social mores, particularly since the Victorian age, prompt anxiety about the public appearance of bodily waste. In some cases, our decay threatens our actual physical well-being. Improperly disposed of waste and corpses can lead to illness. Mucus, saliva, and blood can transmit disease. In other cases, decay threatens our figurative psychic well-being. Lost hair, skin, and blood remind us of our body's inevitable decline. Our anxiety about our decaying bodies, both rational and irrational, prompts us, in some rather irrational ways, to fear decay itself. We banish it to avoid discomfort, and when we do come into contact with it — particularly when that contact is not planned — we feel revulsion. It sickens us. We vomit as if to force all decay in our body out and away.

While our rational fears of the corpse do provoke us to cast it away, it is our irrational fears that trigger abjection. Rationally speaking, the corpse does forewarn us of our own inevitable death, and it is true that close association with a corpse can lead to our more immediate death. Vomiting, however, will not necessarily protect us from either the reality of our eventual death, nor the contagions carried in the corpse. Vomiting is an irrational response. We fear the corpse in ways more profound than reason can explain. The physical, nameable object that is a corpse must be banished because it has the power to harm our bodies, but that psychic, indefinable thing that a corpse suggests, the abject, must be banished because it has the power to define us, to alter our identities, to destroy our subjectivity.

The corpse, through Kristeva's exemplar of the abject, is perhaps too literal a thing, too much an *object*, to fully explicate the power that the abject holds over us. The corpse not only justifies— through its literal potential to destroy life — the need to banish it, but also— through the rituals of burial, cremation, and so forth — allows us to imagine the successful execution of that banishment. The abject, the irrational fear we attach to the corpse, is never so easily identified or disposed of. Kristeva writes: "Getting rid of it [the abject] is out of the question [...] one does not get rid of the impure; one can, however, bring it into being a second time, and differently from the original impurity."[6] We cannot name the abject, concretely define it, nor ever fully rid ourselves of it; yet, we must grapple with it for we face it, we are it, everyday of our existence. We re-present it, re-stage it, and thereby attempt, if only for a fleeting moment, to own it; and in some cases we simply re-present it to celebrate it.

Abject Queers

I have presented the abject thus far, via Kristeva, as a psychic negative, a necessary evil we must negotiate only because we can never wholly eradicate it. Yet, there are those who champion the abject, who identify *as* abject, embracing the instability inherent in such an identity claim. Not all wish to purge themselves of the abject, to jettison their own abjectness. Some embrace the freedom that comes from living at the border between a stable identity and the abyss of unknowns. They proudly wield the power to upend norms and destabilize the identities of those who dare deny abject existence. These are those we call queer.

By queer I do not refer simply to those who engage in same-sex sexual relations. While some homosexuals (gays, lesbians, and bisexuals) are queer, and many queers participate in same-sex sexual relations, to be one is not necessarily to be the other. Homosexuals, on the one hand, serve as the *other* to heterosexual normativity. Homosexuals seek acceptance within heterosexual society. Queers, on the other hand, do not ask for acceptance. More so, they make their acceptance impossible. By never clearly anchoring their identities to any one subject position, by refusing to participate in the subject/object dichotomy, queers thwart attempts to categorize them, elude any subject's efforts to label them as *other*. Queers are those who exist outside of the binary of subject and object that regulates normative culture. Queer, in this text, refers to the abject of society, to those who, more specifically, embrace their abject sexuality.

Queers have power by virtue of their distinction from the homosexual, heterosexuality's *other*. Queer identity, unlike the homosexual object, does not depend on the stability of the subject, and, unlike the normative subject, queers have no stable identity to lose. Their occupation of this abject realm makes queers dangerous, effectively and beautifully dangerous. Their self-proclaimed queerness celebrates and highlights the power of abjection's persistence.

Body Worlds and Abject Conquest

A sign posted near the ticket booth at Charlotte, North Carolina's Discovery Place explains that only one hundred people every fifteen minutes will be allowed admittance to the Body Worlds' exhibit. The suggestion that four-hundred people might arrive during any given hour to see a display of plastinated corpses seems absurd, yet in some locations so many people have waited in line to see Body Worlds that promoters have provided twenty-four hour a day access to the exhibits.[7] The popularity of Body Worlds seems surprising for two reasons: one, the exhibits are billed as educational, and educational exhibits do not typically draw large audiences; two, they are educational exhibits of dead bodies, prototypes of the abject. Yet millions of people have paid to look at exhibits that are clearly oppositional to their own normativity.

According to von Hagens, Body Worlds' popularity stems from centuries of interest in human anatomy. People simply want to know more about their own bodies.[8] However, von Hagens' other statements on the subject suggest Body Worlds' draw is less a testament to humanity's drive for knowledge than it is evidence of our cultural obsession with objectifying the abject, normalizing the queer:

> Beyond their didactic qualities, plastinated specimens also radiate a certain fascination, based primarily on their authenticity. Plastination stops decomposition and dehydration so completely that the insides of bodies cease to be objects of revulsion. Observers are not bothered by any kind of offensive odors.

> Plastination creates beautiful specimens as a sensuous experience that are frozen at a point between death and decay.[9]

The plastinate, while having its educational value, also serves to make the corpse safe for lay audiences because it no longer smells like a corpse, like death and decay. Because the dead body no longer repels us, we are able to approach it rationally and safely, giving our own impending deaths, in von Hagens' words, a "certain proximity to reality, which imbues the image of death with a particular reconciliation."[10] At the same time, we no longer feel revulsion at the thought of our own corporeal decomposition since we can now imagine our bodies in an eternal state of stasis. Von Hagens explains: "Plastination is ... able to satisfy the desire for immortality, which until now has been monopolized by the church, in a way that is commensurate with our times."[11] We feel connected to the plastinate for we see ourselves in it, but we only see ourselves in the plastinate because the plastination process has made the body just enough not like us that we can safely contemplate our relationship to it.

This can be most clearly demonstrated by the displays of "whole body plastinates." Commentators have used a polarity of extreme adjectives to describe the whole body plastinate: from "distasteful" and "immoral" to "fascinating and beautiful" and "a valuable lesson."[12] Such descriptions suggest the profound, if not shocking, power of the whole body plastinate on its audiences; yet, the most profound and shocking thing about the whole body plastinate is its lack of any profoundly shocking qualities. Von Hagens has nullified any potential to offend audiences by removing anything abnormal, or corpse-like, from his whole body plastinates. Whole body plastinates are, in a sense, queerly normal.

Take for example the whole body plastinate *The Winged Man*, apparently named for the sheets of skin that radiate from both sides of his body. Such a design concept certainly has the potential to repulse audiences, but *The Winged Man* is not a putrid, decaying corpse that reminds audiences of their own future putrification. Instead, he is a comical plastinate, too silly to scare anyone. His skin has been peeled away from his body and head to reveal everything underneath. The most prominent sheets of skin jut out from his hips and head. They look less like wings and more like big hair and a flowing skirt. Fortunately, this plastinate is given a large, masculine Fedora to wear which counteracts the feminizing effects of the skin sheets. His knees are bent just slightly. His arms are raised at the side of his body with his elbows bent at a forty-five degree angle near his hip skin. His open-palmed hands rest parallel with his shoulders. The resulting effect, especially considering the Fedora, is a plastinate that looks like a cross between Jimmy Durante and Freddy Krueger of the *Nightmare on Elm Street* series. He is scary, like Freddy, but funny, like Jimmy.

The Winged Man's aesthetic presentation seems incongruent with the educational aims named in the statement of purpose on the Body Worlds website.

Even the poster alongside him describes him as a tool of science. The flayed skin, the poster explains, is meant to show what a body looks like without that outer casing. The sheets of skin remain attached at the sides to show the skin's relationship to the body it once covered. The position of the arms and legs demonstrates the functionality of the muscles and tendons. Every element of *The Winged Man*'s design has a scientific, educational purpose — except the hat, which, apparently, is just cute.

Cuteness and anything else that might be aesthetically pleasing is an important component of all of the whole body plastinates in the exhibit. While education is, according to von Hagens, the primary goal of Body Worlds, the didactic function of the plastinate need not trump its aesthetic value. As von Hagens explains: "These gestalt [whole body] plastinates do not serve any artistic purpose; however, they can certainly comply with artistic criteria. Like works of art, well-executed gestalt plastinates possess aesthetic appeal and emotional worth. This is expressed, inter alia, in that half of all visitors have also come to the exhibition because of its aesthetic appeal."[13]

Not only does the emphasis on aesthetics not interfere with the educational function of the plastinates, it enhances it. As von Hagens explains:

> The lifelike poses of gestalt plastinates are so similar to the living that viewers can actually recognize and even feel their own corporeality and can identify with it. Gestalt plastinates are not objects of mourning; they are instructional specimens. Mourning would interfere with learning; our thoughts would digress. Consequently, I have attempted to make gestalt plastinates appear as lifelike as possible. Freed of the stigma of revulsion, such vital, holistic anatomy thus becomes feasible, with which viewers can be fascinated by its authenticity.[14]

In other words, von Hagens designs his plastinates to look enough like us— the still living — that we can relate to, identify with, and not fear the object before us. At the same time plastinates retain enough differences that we cannot empathize with them so much that we begin to see too much of ourselves in them and again fear them as our abject. This detachment allows us to look at the object before us as just that — an object, the *other* to our subjectivity. Von Hagens distances the corpse from all things corpse-like, simultaneously humanizing and objectifying the corpse in a way that wholly removes it from its abject status and renders it no longer a corpse.

As Megan Stern argues in her essay, "Dystopian Anxieties versus Utopian Ideals: Medicine from *Frankenstein* to *The Visible Human Project* and *Body Worlds*," von Hagens neutralizes abject responses to the dead body by aestheticizing it: "The abject status of dead flesh, especially that which has been dissected, relates to its excessive or grotesque quality. Decaying and leaking, it threatens to invade the realm of the living, hence the need to keep it at bay, both through horrified rejection and medical sanitization. But plastination stabilizes the body permanently, neutralizing any whiff of decomposition and leaving behind bodies that are extraordinarily undisturbing."[15]

Von Hagens takes what is most abject to normative society and attaches normative identity markers to it. He does so most effectively through his use of heteronormative gender tropes. In her essay, "Touching the Corpse," Uli Linke points to the privileging of male over female genitalia in the Body Worlds exhibits. She writes: "These plastinated and often elongated or erect penises, complete with scrota, are displayed in separate showcases. By contrast women's reproductive organs, that is, vagina and uterus with ovaries, are exhibited with malignancies, tumors, deformities and cysts." This presentation of female organs, claims Linke, "pathologizes the female body."[16]

In addition to the emphasis on diseased female organs, few of the full body plastinates are female, and most of those emphasize reproduction. One shows a pregnant female with a fetus inside of her, another reveals the make-up of the female reproductive system, and the others show the plastinates doing normatively feminine tasks. In one display in particular, *The Figure Skating Pair with Woman Lifted*, a male plastinate hoists his female partner over his head. Her back arches over his extended arm and hand. This display claims to demonstrate the male's muscular strength and control. The female functions as an object for demonstration.

It is interesting to note that the female in *The Figure Skating Pair* display is one of the few female plastinates in which the labia are still intact. While the male plastinates almost invariably have their penises intact, as Linke also notes, almost all female plastinates are missing their labia and clitorises. The female plastinates, however, all have visibly erect nipples.[17] Linke argues that these biased impositions of gender norms on male and female plastinates obscure the audience's ability to critically reflect on von Hagens' "fabrication of biomedical truths." She writes: "By way of these biological phantasms, culturally specific notions of sexuality are reified as natural or authentic anatomy."[18] By using socially constructed notions of gender identity to animate his plastinates, von Hagens creates a linkage between scientific "fact" and gender norms.

Consider, for example, the pregnant female plastinate. She lies on her right side, propped up on her elbow. Her back arches, pushing forward her pregnant belly. An incision running from just below her sternum down to her vaginal opening reveals a fully formed fetus crouching inside her uterus, which is also sliced open. The top of the incision creates an arrow that points to her chest and draws the viewer's eyes to her breasts with their oddly protruding nipples. Her left arm bends above her with her left hand resting on the back of her downward turned head. Her eyes also look down and away from the audience. The bottom of the incision points to her vaginal opening where, unlike the prominently displayed breasts above, the arrow created here points to nothing. Her clitoris and labia have been removed, leaving behind a slice of vacant space that, with her legs spread just slightly, seems to highlight the exit route for the fetus inside.

Her pose is that of a reclining nude from the classical era of Western art, reminiscent of a work by Titian or Giorgione. Her arched back emphasizes her supple curves. Her shameless nudity invites the viewer to consume her body. Her downward gaze eliminates any question of her resistance or agency. She is the ideal object of the masculine gaze. Unlike in a Titian, however, her sexuality is downplayed. By removing her clitoris and labia and emphasizing her nipples and the birth canal, von Hagens creates a woman whose sexual purpose is wholly reduced to her reproductive function.

The following is his defense of his presentational choices of this controversial display:

> It was difficult to position this gestalt plastinate: the body should not be made to appear lifeless and should permit an optimum view of the fetus in the uterus, while at the same time showing proper respect for this double tragedy. The physical discomfort of an advanced state of pregnancy made the reclining pose obvious. In this connection, this pose, in which the body is lying upright on its side, has aided in providing the desired lifelikeness as well as the necessary tension. The head has been tilted to the side with the eyes closed, i.e., consciously away from the viewer, and the hand of the free arm has been laid on the back of the head, as with a headache. In this way, the body language of this pose fundamentally differs from a lascivious provocation, in which the head would be coquettishly thrown back and the eyes would be enticingly trained on the viewer.[19]

Von Hagens justifies his aesthetic choices through science. The back is arched to "permit an optimum view of the fetus in the uterus." She assumes a reclining position because of the "physical discomfort of an advanced state of pregnancy." She looks demurely away from audiences to avoid a "lascivious provocation." Oh, and her hand is on the back of her head because she has a headache. His justification, as absurd as it is, reveals the gender biases inherent in his staging of plastinates. That a pregnant woman feels discomfort in advanced stages of pregnancy is a potential scientific fact. Linking that to a pregnant woman propping herself up on one arm with her unsupported back lurching forward only serves to justify a socially constructed gender norm. A hand on the head can serve as an indicator of a headache, but that hand usually goes to the front of the head and the arm supporting that hand is rarely so taut. Placing the plastinate on her side with her stomach pressed forward does highlight the view of the fetus inside the uterus, but what purpose do the erect nipples and missing genitalia serve? Finally, having her eyes point away from viewers does not desexualize the plastinate; it dis-empowers her; it makes her stereotypically, passively feminine.

Removing the labia and the clitorises of the female plastinates couches their sexual anatomy in reproductive terms only. The male plastinates, as stated earlier, almost always have fully intact gonads and penises. Even when the plastinate on display has nothing to do with male reproduction — i.e. *Winged Man*— his genitals are still prominently displayed and unaltered. In the few cases where the female partner's labia and clitoris remain intact, such as in *The Figure Skating Pair*, her passive, feminine role is clearly defined and her sexuality is clearly juxtaposed with a male plastinate's dominant sexuality.

By objectifying the abject corpse, von Hagens gains control over the meaning-making potential of his object — the plastinate, previously a corpse. Once static, no longer illusive in its abjection, the identity of the plastinate becomes subject to the anatomist who designs it. The corpse loses its ability to terrify, and thereby control, the subject. Once neutralized through objectification, the plastinate becomes a kind of blank canvas upon which normative subjects can paint their own reflections. Namely, imposing normative gender codes on the plastinate affirms the stability of the gender binary. The scientifically justified presentations stage the socially constructed differences between men and women as part of our inherent biology. Von Hagens placates the anxieties of his audiences by confirming the inevitable stability of the ever-pervasive gender binary. In other words, he allows the plastinate to function as the *other* to the subjectivities in von Hagens' audiences.

Conclusion: Abject Triumph

Full body plastinates, such as *The Winged Man*, embody the essence that is Body Worlds: a little silly and a little creepy but not enough of either to be anything except sensationalist. When placed in the context of the exhibit as a whole, *Winged Man* highlights the absence of any corpse-like qualities present in any of the plastinates. He lacks the markers of morbidity and is more likely to provoke a chuckle than a fainting spell. The plastinates on display in Body Worlds are all too strange in a normative kind of way to help audiences, as von Hagens claims, to confront their irrational fear of the human corpse in order to learn more about their bodies, "develop anatomical pride in our bodies," and, thus, "emancipate the body" from society's repressive relationship with it.[20]

While Stern and others[21] have argued that Body Worlds presents the abject, I argue that the process of plastination and the normative staging of the plastinated specimen change the dead body from an abject corpse into a safely *othered* object, a failed subject. The plastinate retains none of the power of the corpse's abjection because the process of decay has been halted. As a result, von Hagens' plastinates do not challenge abjection. They circumvent it. Von Hagens fails in his quest to nullify the abject power of the *actual* corpse because the object he presents in his exhibits is no longer a corpse but a safely contained object called a plastinate. Visitors to Body Worlds are not cured of their irrational fear of the corpse, nor will they become comfortable confronting the reality of their own inevitable decay. Visitors to Body Worlds only find temporary solace from the horror and repulsion that a rotting corpse provides.

Notes

1. von Hagens, "On Gruesome Corpses," 260, 280.
2. Kristeva, *Powers of Horror*, 1.
3. Ibid., 1–2.
4. Ibid., 3.
5. Ibid.
6. Ibid., 28.
7. Grayson, "Science Museum's Body Worlds Exhibit."
8. von Hagens, "On Gruesome Corpses," 277.
9. von Hagens, "Anatomy and Plastination," 20.
10. von Hagens, "On Gruesome Corpses," 272.
11. Ibid.
12. See Youngs, "Corpse Show Leaves Sour Taste"; Skelton, "Catholic Officials"; Williams, "The science exhibit"; and Brecht, "Body Worlds."
13. von Hagens, "On Gruesome Corpses," 272.
14. Ibid., 262.
15. Stern, "Dystopian Anxieties versus Utopian Ideals," 79.
16. Linke, "Touching the Corpse," 17.
17. Ibid.
18. Ibid.
19. von Hagens, "On Gruesome Corpses," 264.
20. Ibid., 268.
21. See also Muller, "Abject d'Art"; Lewis, "Body and Soul."

Bibliography

Brecht, Philip. "Body Worlds: The Anatomical Exhibition of Real Human Beings." *Primary Health Care* 12, no. 5 (2002): 10.

Grayson, Kathy. "Science Museum's Body Worlds exhibit open 24 hours." *Minneapolis/St. Paul Business Journal*, November 28, 2006. http://twincities.bizjournals.com/twincities/stories/2006/11/27/daily16.html (accessed January 13, 2008).

Kristeva, Julia. *Powers of Horror: An Essay on Abjection*. Trans. Leon S. Roudiez. New York: Columbia University Press, 1982.

Lewis, Michael J. "Body and Soul." *Commentary Magazine*, January (2007), 29–32.

Linke, Uli. "Touching the Corpse: The Unmaking of Memory in the Body Museum." *Anthropology Today* 21, no. 5 (2005): 13–19.

Muller, Vivienne. "Abject d'Art." *M/C Journal* 9, no. 5 (2006). http://journal.media-culture.org.au/0610/04-muller.php

Skelton, Chad. "Catholic Officials Boycott Body Worlds Exhibit," *Vancouver Sun*, September 15, 2006.

Stern, Megan. "Dystopian Anxieties versus Utopian Ideals: Medicine from Frankenstein to the *Visible Human Project* and *Body Worlds*." *Science as Culture* 15, no. 1 (2006): 61–84.

von Hagens, Gunther. "Anatomy and Plastination." In *Gunther von Hagens'*, 2–36.

_____. "On Gruesome Corpses, Gestalt Plastinates and Mandatory Interment." In *Brave New Body Worlds — The Question of the Exhibition*, edited by Franz J. Wetz and Brigette Tag, 260–282. Stuttgart: Klett-Cotta Verlang, 2001.

_____, and Angelina Whalley, eds. *Gunther von Hagens' Body Worlds: The Anatomical Exhibition of Real Human Bodies*. Heidelberg: Institut für Plastination, 2004.

Williams, Kevin. "The science exhibit 'Body Worlds' has generated controversy every stop of its tour for being too realistic, even gross. But we've seen it, and we think it's amazing. So have no fear." *Chicago Tribune*, February 18, 2005. http://www.chicagotribune.com/features/lifestyle/health/chi-0502180429feb18,1,3729787.story?coll=chi-health-utl (accessed January 13, 2008).

Youngs, Ian. "Corpse Show Leaves Sour Taste." *BBC News Online*, March 22, 2002. http://news.bbc.co.uk/1/low/entertainment/arts/1888283.stm (accessed February 12, 2008).

Section Five

Aesthetic and Instructive but Not Morally Offensive

Anatomy Without Integrity

Ruth Levy Guyer

Integrity: n. Adherence to ethical principles[1]

Curators at museums of history, natural history, science, and medicine have long grappled with how most respectfully to acquire, handle, and display bodies and their parts— skeletons, tissue specimens, mummies, fetuses in bottles, and others. Formulaic approaches have never existed, because cultures, religions, societies, and individuals have differing perceptions of death, its rituals, and the integrity and dignity of the human body. Still, knowing the provenance of each body specimen seemed to matter.

Then, along came Body Worlds. Some years ago, Gunther von Hagens, the curator of Body Worlds, moved his plastination enterprise from Germany to China and brought his exhibits to North America because of "ongoing legal investigations in Germany."[2] In 2002, Russian prosecutors charged medical officials in Siberia with having shipped fifty-six bodies and more than 400 brain parts to von Hagens in Germany without obtaining permission from the families of the dead.[3] A few months later, thirty-two more bodies from Siberia were intercepted by the police.[4] In 2004, media reports suggested that von Hagens was using executed Chinese prisoners for plastination; again there was no proof, but von Hagens ended up cremating seven corpses, all with head injuries, leading to speculation that those individuals had died by bullets to the head, the Chinese method of execution.[5]

These suspicions and allegations are troubling, no one is at all clear where the bodies in Body Worlds came from, and von Hagens, the synecdochically self-described "brains" of the exhibit, isn't telling. Yet museums around the country— their curators and their review panels—continue to give the issue of the bodies' provenance an easy "pass." And the gawking public and much of the media have done the same. The open solicitations at the ends of the exhibits of Body Worlds asking viewers to sign up to be body donors themselves probably have contributed to the conclusion that the bodies on display were those of willing donors. There is, however, not a shred of evidence for this presented with the exhibit.

Von Hagens has not been convicted of any crimes but his lack of candor about the bodies is curious. He actively obfuscates. He readily provided evidence of a paper trail between signed consent forms and death certificates, but then he would not or could not show that the bodies in his exhibit were related to those documents; thus the forms and certificates were worthless. He claimed that, for privacy reasons, he had created a firewall between the bodies and their identities.[6] Through such actions, von Hagens dismisses, mocks, and trivializes the very important concept of informed consent, ignoring all of the ethical guidelines that were hammered out in the twentieth century with respect to the appropriate treatment of human subjects in research and medicine.

Consent that is truly informed is not easy to verify. Few researchers actively test those who volunteer in order to make sure that they understand what they have volunteered to do. Still, sometimes it has been very easy to ascertain that a person has wanted to be part of an anatomy exhibit.

For example, one of the most fervent body-parts donors ever was surely Major General Daniel E. Sickles, whose right leg bones are on display at the National Museum of Health and Medicine outside Washington, D.C. Sickles' leg was shattered by a twelve-pound cannonball on the second day of the Battle of Gettysburg, and it was quickly amputated. Sickles purchased a mini-coffin, placed his leg in it, and immediately shipped the package to the museum with this note: "With the compliments of Major General D.E.S." In the display case, along with Sickles' leg, are a cannonball and a set of photographs that show the Major General at different times, standing in front of the display case, paying homage to his lost limb, often surrounded by friends at the museum.

Another similarly eager body donor was Jeremy Bentham, the father of utilitarian philosophy. Utilitarian to the end, Bentham specified that, when he died, he wanted his body to be publicly dissected and then reassembled into a true-to-life instructional effigy, his "auto-icon." "This my will and special request make," wrote Bentham, "that mankind may reap some small benefit in and by my decease."[7] Bentham's physician and friend performed the autopsy in London on a day of wild storms in 1832 and described what he was doing to the audience "with a clear unfaltering voice, but with a face as white as that of the dead philosopher before him."[8] Bentham's skeleton was then padded with cotton wool, straw, hay, paper ribbon, and other stuffing materials and dressed. A wax likeness of Bentham's head was prepared and placed on top of his body. Glass eyes that Bentham had purchased animate his wax face. Bentham's auto-icon still sits in a mahogany and plate glass case in University College London.[9] (His actual head was damaged during the preservation process and is in a vault.) The enduring "utility" of the auto-icon seems to be largely that of a target for sophomoric pranks and as the subject of occasional academic inquiries and musings.

At the other end of the spectrum from Bentham and Sickles was the completely demurring body "donor," Charles O'Brien, the eighteenth-century Irish

Giant whose imposing skeleton is part of the Hunterian collection at London's Royal College of Surgeons. (He is also referred to as Charles Byrne.)[10] O'Brien was a local wonderment in London, renowned for lighting his pipes with flames from street lamps. He was almost eight feet tall, a victim of acromegaly or giantism. He spent much of his life on display at London's Cox Museum, where visitors paid half a crown to gawk at him. He dreaded the possibility that his skeleton would be exhibited after his death — he did not want to live on in perpetuity next to criminals and other low-lifes, the stock-in-trade sources of skeletons for anatomists in the 1900s.

When O'Brien was twenty-two and realized that he was dying (of heart and other organ damage caused by giantism and alcoholism), he arranged with local fishermen for an at-sea burial in a lead-lined coffin. But John Hunter, the Surgeon Extraordinary to the King of England, bribed the undertakers to ignore O'Brien's arrangements and, instead, turn the cadaver over to him, which they did. Hunter boiled O'Brien's skeleton. This stripped the flesh from the bones. Then he hid the bones in his basement for several years. Eventually, he assembled the skeleton and put it on display. When I visited the Hunterian Collection, some 300 years after O'Brien's death, O'Brien's skeleton was still on display.

I was at first awestruck when I encountered O'Brien's massive skeleton. For starters, my feet and his rested on the same floor yet my eyes were level with his pelvic bone.[11] But then I was appalled to read, right next to the skeleton, the account of O'Brien's directives about his burial preferences. Not only had Hunter ignored O'Brien's clearly expressed wishes to rest in peace in a watery grave, but so had all of the museum's curators for three centuries. How extraordinary that no one has honored O'Brien's request in all this time, especially because today we know of so many examples of family members and communities trying to reclaim the bones of their ancestors or the bodies of the survivors of genocides and violence in order to properly bury them.[12]

In this contemporary climate of enhanced awareness, it is also surprising that curators at each museum that has hosted Body Worlds continue to buy into von Hagens' claim not to worry, that all is well with the bodies in his exhibits. I saw, first hand, the total absence of serious deliberation about possible ethical problems with Body Worlds when I was invited to participate in an ethics panel that was reviewing the exhibit for the Franklin Institute in Philadelphia. The panel consisted of about twenty-five people. All but two were men, many of the men were clergy, and the other woman was very old and had been a nurse. The museum staff sent each of us a booklet ahead of time about the exhibit and then, when our panel met, presented their case for hosting Body Worlds.[13] I was the only person who voiced any objections and concerns. When I commented that the consent form associated with the exhibit was simplistic and inadequate, one of the other panelists said he'd like to see the form. (It was in the packet of materials that he and all of us had received ahead of time.) When

I mentioned something about the partially dissected pregnant woman and her partially dissected fetus that are in the exhibit, one person expressed concern that "the fetus did not give consent." I asked what plan existed for disposing of any of the bodies that might decay or fall apart, and that also interested the panelists. In the end, though, there was no authentic ethics review, not a chance (or any expectation) that the museum would not host this money-making show.

The unquestioning acceptance of von Hagens' claim that he is following in the giant footsteps of Andreas Vesalius, the Renaissance anatomist, is also interesting. Vesalius included engravings of posed human skeletons and partially dissected bodies framed by scenes of the Padua countryside in his classic anatomy text *De Humani Corporis Fabrica* (*On the Fabric of the Human Body*). He meant the images to be both anatomically and philosophically instructive. One skeleton stands at a tomb fingering a skull and is said to be contemplating its own mortality. Another leans on a shovel, appearing to have just dug his own grave. Yet another is bent over in obvious grief. What distinguishes Vesalius' skeletons and corpses from those of Body Worlds, though, is their form: Vesalius immortalized bodies in engravings, not in the flesh.

In the years since Vesalius' time, the human body has not evolved very much. But surely our awareness of our ethical responsibilities to human bodies should have. Four centuries have passed since Vesalius was alive. One of these was the twentieth century, during which landmark events forever affected our understanding of the importance of universal human rights and the responsibilities and obligations of medical researchers to those who "volunteer" to donate their bodies to medicine and scientific research. We understand the importance of respecting the autonomy of others, particularly as it relates to their participation in medical research. This respect, for humans and their bodies, includes both the period when people are alive and after their deaths.

The first significant step in directing our attention to the need for respecting the autonomy of persons was the trial of Nazi doctors at Nuremberg after World War II, which led to the development of the Nuremberg Code.[14] The Code begins, "The voluntary consent of the human subject is absolutely essential." It goes on to point out that "[t]he duty and responsibility for ascertaining the quality of the consent rests upon each individual who initiates, directs or engages in the experiment. It is a personal duty and responsibility which may not be delegated to another with impunity."

All who have been asked to review the Body Worlds exhibit and all who visit museums housing Body Worlds have obligations not to just passively stare at what is on display but to inquire actively about the exhibits, demanding to know the full stories, including the back stories, of the bodies. Ethical responsibility has no statute-of-limitations. A relevant illustration of how individuals have drawn attention to ethical violations long after they occurred is found in what has been happening with Eduard Pernkopf's *Topographische Anatomie des Menschen*, an anatomy atlas that has been in wide use by physicians and surgeons since the Second World War. About twenty years ago, doctors first suspected that drawings in the atlas were tainted; they initiated an investigation, and their close scrutiny led to important conversations and deliberations about who the models were for the drawings and what lessons the users of the atlas could learn from their findings. Pernkopf's atlas includes approximately 800 meticulously drawn watercolors. Certain physical features depicted in the drawings— sparse hair, wasted appearances (possible signs of malnutrition), a baby with an attached umbilical cord — suggested to United States and Canadian physicians that the models for the atlas may have been victims of Nazi atrocities.[15] Thus, in 1995, the University of Vienna, where the atlas had been produced, undertook a formal investigation into where the university's Anatomy Institute got the specimens that were in its anatomy collection. Their 1998 report documented that 1,377 bodies of the 4,000 bodies in the collection were those of executed political prisoners. Most died on the guillotine; others were shot for crimes ranging from high treason to pick-pocketing. Records for the rest of the bodies— their provenances, who they were, how they died— could not be determined. All that was known was that they were sent to the institute for "research and teaching."[16]

Pernkopf, an ardent Nazi, became the director of the Anatomy Institute in 1933. He hired National Socialist Party members and sympathizers as artists to prepare drawings for the atlas. Their work spanned the 1930s and 1940s and continued into the 1950s. Sometimes the artists inserted Nazi symbols— swastikas, double lightning bolts— into their signatures. In later editions these were airbrushed out, although some tell-tale signs remain. The signature of Karl Entresser, for example, appears in the third (1987) edition of the atlas in two forms: in some drawings "ss" has normal curves; in others, those two letters are simply two slashes— "//" — the residue of the lightning bolts.

As physicians, medical librarians, institutions, and others have learned more about the Pernkopf Atlas, they have considered carefully what actions to take. What would be the best, most respectful, and fitting memorial to the dead? Should they destroy the atlas, even bury it? Might continued use of the atlas suggest that atrocities can simply be wiped clean and thus give license to other researchers contemplating unethical or inhumane experiments? A few medical libraries have removed the atlas from their collections. But most have chosen to house the atlas among their special collections. There they feel they can balance access to the atlas with an opportunity to draw attention to the violations of medical ethics associated with the atlas' production. They argue that continued use of the atlas would help ensure that the people whom the Nazis chose to exterminate are, instead, immortalized. One example of this is explained by United States physician, Howard Spiro, who wrote in a 1998 article that continuing to have the atlas available could teach both medical and moral lessons.[17] He noted that stopping publication or preventing use of the atlas would resemble Nazi book burning. He suggested that the rubbed out Nazi symbols be restored to the artists' signatures. "The crimes of the Nazi doctors [can] teach us what we physicians must not do." And, he wrote, "The brilliant depictions of Pernkopf's atlas are transfiguring; in each, I hear the scream of a person."[18]

The open and honest discussions that have taken place around the Pernkopf inquiry seem to me to be just what should be going on around the Body Worlds exhibit. Any museum that is considering installing the show or already has should thoroughly investigate the thorny issue of the provenance of the bodies. Suppose the bodies are those of executed prisoners or bodies "pirated" (as one report characterized the situation) [19] from medical and psychiatric institutes? Surely those individuals were not authentic volunteer body donors. Should such bodies be taken off display or should they remain on display but exhibited with clear narratives that draw attention to exactly what the story is? Museums that have installed or plan to install Body Worlds should insist on seeing the paper trails for each body in the show and should openly display the paper trail for each. For those for which the trail is cold, museums should reconsider what to do. If bodies are those of unwilling or improperly "consented" donors, should they immediately be taken off display and properly buried? Or should information about their plights at least be offered to museum visitors?

Certainly, the least satisfactory solution is to continue to ignore the provenance issue, which is what is now being done. Visitors must learn the complete story, even if the story is unpleasant (as many suspect) or even if some details remain murky. Many visitors to Body Worlds probably do not, on their own, think about the origins of the bodies, but that is one lesson that museums could be teaching to the viewers as a central issue of this exhibit. (A man in line next to me at Body Worlds had no idea that he was even looking at real bodies; he thought the bodies were plastic models, because the exhibit is so thoroughly disconnected from, and unwilling to mention, its true subject: death.)

Making anatomy democratic is a stated goal of Body Worlds. Democratization was also a goal of Vesalius, Bentham, and many others. Human anatomy, once the exclusive province of doctors and other medical personnel, is increasingly available in the public domain, thanks largely to the internet. Today, anyone with access to a computer can see every organ, cell, tissue, and fiber of the human body and learn about them — though not smell or touch them — by pulling up, for example, the homepage of the Visible Human Project.[20] This website came online in 1994 and it too has met with ethical objections, although at first, acclaim for the Visible Human Project was widespread. Two people — a man and a woman — were frozen solid at their deaths, quartered with an electric saw, and then sliced, like huge salamis, into some 1,800 ultra-thin slices each. Histologists photographed every slice, reassembled them, and put the images online. Viewers are able to situate a virtual rectangle over a body part, then click to expand or narrow the focus, and home in on a structure. The online images help surgeons and other physicians in ways that, in the past, only anatomy texts were able to do. In fact, the Visible Human Project and other web-based, high-tech, computerized resources are changing the way that some medical schools are teaching anatomy.

Users initially applauded the Visible Man's generous donation to medical science. But then they discovered that the Visible Man had been a ruthless killer who had bludgeoned his victim, an old man, to death. The accolades stopped. The questioning began. Why choose such a man to serve as the standard for all men? Why confer a kind of immortality on the body of a felon? The researchers who developed the Visible Human Project described their need for a physically exquisite, healthy corpse, one not compromised by disease or injury or old age. A death-row prisoner seemed ideal — the person could be executed at a known time, quickly frozen, and immediately processed.

How do we know that this prisoner, in fact, did make a free and autonomous choice to be the Visible Human? Prisoners lose many rights, and it is not at all clear that the Visible Man's consent was fully informed. How much could he have known about the novel procedures that would be used and their consequences, for which there was absolutely no precedent? Could he (and later, the Visible Woman) have understood that his body would end up being so thoroughly, lastingly, and globally public and visible? Was the Visible Man an enthusiast, like Sickles and Bentham? Or was he, like O'Brien, an unwitting and unwilling "volunteer?"

And what about the Visible Woman? Little is known publicly about her, except that no one came to claim her body when she died. The Visible Woman died of a heart attack at age 59, which raises the question: What happened to the "healthy" standard that the researchers claimed was essential? Was the main criterion for choosing the Visible Woman that she was someone who had been effectively invisible in life?

Those who willingly donate their bodies to research or to exhibits, as Sickles and Bentham did, are ready to contribute to advances in medicine and in that way hope to make lasting contributions. Others, like O'Brien, ask simply to be buried — and others ask to be cremated — when they die. All of these are valid requests that should be honored by the living. Those who use cadavers in their work — as von Hagens has in plastination, as other museum curators have, as the online resources have — have solemn obligations to provide complete and transparent information to both the potential donors and to those who view the bodies and body parts after the deaths. Those who look at and learn from these exhibits and resources— all of us— also have responsibilities to the dead, to demand that ethical standards are fully met.

I have reflected previously on the relevance of Lewis Thomas' description of the poignant burial rituals of elephants to what I find disturbing about Body Worlds.[21] Thomas wrote in his essay "Death in the Open" that "If an elephant missteps and dies in an open place, the herd will not leave him there; the others will pick him up and carry the body from place to place, finally putting it down in some inexplicably suitable location. When elephants encounter the skeleton of an elephant out in the open, they methodically take up each of the bones and distribute them, in a ponderous ceremony, over neighboring acres."[22]

The elephants, through their basic animal instincts, remember what we humans—with our so-called highly evolved brains—are overlooking, forgetting, or actively disregarding—that the dead deserve to be treated with dignity. It is never too late for us to remember and fulfill our ethical obligations to one another.

Notes

1. Random House Webster's Unabridged, 2nd ed., s.v. "Integrity."
2. Linke, "Art for Art's Sake," 23.
3. "Russians Charged."
4. Rowland, "Scientist Denies Link to 'Snatched' Corpses."
5. Nickerson, "Traveling Cadaver Show," A1.
6. Tanassi, "Responsibility and Provenance," 36; Ulaby, "Cadaver Exhibits."
7. Marmoy, "The 'Auto-Icon,'" 77.
8. Ibid.
9. University College London.
10. Richardson, "A Potted History," 935.
11. Guyer, "Mind's Eye," 108.
12. Harper, *Give Me My*.
13. California Science Center, Gunther von Hagens' Body Worlds.
14. The Nuremberg Code online.
15. Israel and Seidelman, "Nazi Origins," 1633.
16. Hubbard, "Eduard Pernkopf's Atlas," 207.
17. Spiro, "The Silence of Words," 183.
18. Ibid.
19. Nickerson, A1.
20. National Library of Medicine.
21. Guyer, "Metamorphosis," 30; "A Bioethicist."
22. Thomas, *The Lives of a Cell*, 113–116.

Bibliography

California Science Center. "Background Information." Exhibition Pamphlet. Gunther von Hagens' Body Worlds: The Anatomical Exhibition of Real Human Bodies. Los Angeles: 2005.

Guyer, Ruth Levy. "A Bioethicist Takes a Peek at Body Worlds." *Weekend All Things Considered*. National Public Radio, August 12, 2006. http://www.npr.org/templates/story/story.php?storyId=5640183 (accessed August 26, 2007).

_____. "Metamorphosis: Beautiful Education to Smarmy Edutainment." *American Journal of Bioethics* 7, no. 4 (2007): 30.

_____. "Mind's Eye: Giants in the Earth." *Potomac Review*, Spring 2000, 108.

Harper, Kenn. *Give Me My Father's Body: The Life of Minik, the New York Eskimo*. New York: Washington Square Press, 2000.

Hubbard, Chris. "Eduard Pernkopf's Atlas of Topographical and Applied Human Anatomy: The Continuing Ethical Controversy." *Anat Rec* 265 (2001): 207.

Israel, Howard, and William Seidelman. "Nazi Origins of an Anatomy Text: The Pernkopf Atlas." *JAMA* 276, no. 20 (1996): 1633.

Linke, Uli. "Touching the Corpse: the Unmaking of Memory in the Body Museum." *Anthropology Today* 21, no. 5 (2005): 13–19.

Marmoy, C.F.A. "The 'Auto-Icon' of Jeremy Bentham." *Medical History* 2 (1958): 77–86.

National Library of Medicine and National Institutes of Health. "Overview." The Visible Human Project website. Bethesda, MD: 2003. http://www.nlm.nih.gov/research/visible/visible_human.html

Nickerson, Colin. "Traveling Cadaver Show Wants You: Body Exhibit Seeks Donors in Boston." *Boston Globe*, July 9, 2006, A1.

Richardson, Ruth. "A Potted History of Specimen-Taking." *The Lancet*, 355 (2000): 935.

Rowland, Jacky. "Scientist Denies Link to 'Snatched' Corpses." *BBC News*, December 2, 2002. http://news.bbc.co.uk/2/hi/europe/2535471.stm (accessed August 26, 2007).

"Russians Charged Over Body Parts." *BBC News*, July 20, 2002. http://news.bbc.co.uk/2/hi/europe/2140333.stm (accessed August 26, 2007).

Spiro, Howard. "The Silence of Words— Some Thoughts on the Pernkopf Atlas." *Wien Klinische Wochenschr* 110, no. 4–5 (1998): 183.

Tanassi, Lucia M. "Responsibility and Provenance of Human Remains." *American Journal of Bioethics* 7, no. 4 (2007): 36.

Thomas, Lewis. *The Lives of a Cell.* New York: Bantam Books, 1974.

Ulaby, Neda. "Cadaver Exhibits Are Part Science, Part Sideshow." *All Things Considered*, National Public Radio, August 10, 2006. http://www.npr.org/templates/story/story.php?storyId=5553329 (accessed August 26, 2007).

United States Department of Health and Human Services. "Office of Human Subjects Research." The Nuremberg Code. Washington, D.C: 1997. http://www.hhs.gov/ohrp/references/nurcode.htm.

University College London. "The Auto Icon." UCL Bentham Project. London: 1999. http://www.ucl.ac.uk/Bentham-Project/info/jb.htm

Adoration, Veneration, Plastination: Theo-liturgical Reflections

Paul J. Wojda

Life is short, but Plastination will endure (and will educate) for a long time.

— Gunther von Hagens

Mr. McGuire: I just want to say one word to you — just one word.
Ben: Yes sir.
Mr. McGuire: Are you listening?
Ben: Yes I am.
Mr. McGuire: "Plastics."
Ben: Exactly how do you mean?
Mr. McGuire: There's a great future in plastics. Think about it. Will you think about it?
Ben: Yes I will.

—*The Graduate* (dir. Nichols, 1967)

Tantum ergo Sacramentum
veneremur cernui

— St. Thomas Aquinas, *Pange Lingua*

I.

Like all technical achievements, the "Polymer Impregnation of Perishable, Biological Specimens," or as its inventor Gunther von Hagens calls it, "plastination," originated as the solution to a highly circumscribed practical difficulty, in this case the preservation of anatomical specimens for the purpose of medical research and education. As von Hagens recalls, the idea initially occurred to him when, as a young assistant at the University of Heidelberg's Institute of Anatomy, he was confronted for the first time with a human kidney encased in a block of clear polymer. He wondered why the specimen hadn't been preserved from the "inside out," that is, why the plastic, in liquid form, hadn't been perfused throughout the organ.[1] Gripped by the possibility ("I could not get this question out of my mind.") and with the resources of Heidelberg's Institute at his disposal, it was not long before von Hagens had his first successfully "plastinated" organ specimens. Though these initial products were admittedly crude, von Hagens immediately recognized the importance of his discovery for both the field of anatomy and, if his later testimony is to be believed, the rest of his own life.[2]

Over the next two decades (1977–1997) von Hagens refined his technique, the most significant of which allowed him to preserve not just individual organs, but "whole body" specimens, what he calls "Gestalt plastinates,"[3] which could then be further dissected and vertically posed in a variety of educationally useful ways.[4] Meanwhile, patents were continuously acquired, the process exported to more than thirty countries, and in 1993, in Heidelberg, the "Institute for Plastination" founded. Had von Hagens' accomplishments ended there, he could justifiably have anticipated the sort of immortality often accorded to such technical pioneers: the admiration of his peers and educated public; appointment to prestigious academic posts; institutes and endowments bearing his name; perhaps even a Nobel Prize. Instead, von Hagens elected to pursue a different, and decidedly more uncertain fame, for shortly after the founding of his Institute, he introduced his plastinates to the general public, and a good deal of controversy, in the form of Body Worlds: The Anatomical Exhibition of Real Human Bodies.

The exhibit first opened in Tokyo, Japan in 1995, followed quickly by stops in Osaka (1996), and then Mannheim, Germany (1997). Three versions of the exhibit are still touring. To say that these presentations have been popular would be a gross understatement. In just ten years Body Worlds has been seen by well over twenty million people (virtually all of them in the industrial centers of East Asia, Western Europe, and North America), very likely making it the single most viewed museum exhibit of all time. So successful have these exhibits been that von Hagens has opened two "plastination" factories in Dalian, China in order to keep up with demand. Competitor exhibits (e.g., "Bodies," "The Universe Within," "Bodies Revealed," "Body Exploration," "Mysteries of the Human Body," and others) have also appeared, much to the consternation of Body Worlds' promoters.[5] In their imitation, though, they only flatter the original and, of course, von Hagens himself, who with his characteristic black fedora and showman's instincts, very much embraces his role as plastination's chief *impresario.*

Despite its enormous popularity and financial success, Body Worlds has not been able to dispel the aura of controversy that has followed the exhibit since its opening, a controversy caused, in large measure, by the ambiguous origins of at least some of von Hagens' plastinates. In 2004, during one of Body Worlds' last appearances in Germany (as it turns out), the respected newsweekly *Der Spiegel* published reports that at least some of the bodies in the exhibit were those of executed Chinese prisoners, and that others had also been obtained without the consent of the donors or their families. Von Hagens categorically denied, and continues to deny, knowingly using the bodies of executed prisoners, and successfully petitioned for a legal injunction that prohibited *Der Spiegel* from repeating these accusations. The exhibit itself, as well as its accompanying literature, catalog, and website are also at pains to emphasize that all the bodies on display were legally and ethically obtained. Suspicions linger nonetheless, some of them doubtlessly sustained by the confusion of Body Worlds with its competitor exhibits, where such legal and ethical assurances are less in evidence; some by von Hagens' decision to open the two "processing plants" in China, where it appears he has less than full control of operations;[6] and some, as I will suggest below, because moral outrage at the transgression of individual rights (as in the failure to secure the voluntary informed consent of donors) is being made to bear the weight of *other* concerns that Body Worlds raises, which owing to their "religious" character have proven difficult if not impossible to otherwise articulate.

In this essay I attempt to identify with greater clarity and precision the nature of these "religious" concerns from a distinctively Catholic "theo-liturgical" perspective (more on this term shortly). It should be noted at the outset, for the sake of full disclosure, that I do so not only from my perch in the academy, but also as an invited member of an *ad hoc* inter-religious advisory committee assembled in late fall, 2005, by the Science Museum of (Saint Paul) Minnesota in anticipation of the Body Worlds exhibit there in the summer and fall of 2006.* Similar committees had been convened at other North American Body Worlds venues for the purpose of identifying and responding to possible concerns about and even objections to the exhibit coming from local religious communities. Saint Paul and Minneapolis have large Catholic, Jewish, Moslem, and indigenous (Native American, Hmong) populations, and each was represented on the Science Museum committee. A number of committee members, including myself, also contributed to a variety of Science Museum events held in conjunction with the exhibit, e.g., public lectures and panel discussions related to questions about death, funeral/burial rites, and post-mortem human destiny. Members were not compensated for their work on the committee itself. Those who contributed to other events did receive a modest honorarium.

The existence of such committees, in Saint Paul and elsewhere, attests to the recognition on the part of Body Worlds' promoters and its local hosts that

**The views expressed here are, of course, my own, and in no way represent those of either the Science Museum of Minnesota or the advisory committee.*

the exhibit touches specifically religious nerves, even if it is not always clear just what or where those nerves are. Indeed, it would appear that many members of traditional religious communities are not all that clear about these matters either, for when first asked to respond to a recorded preview of the exhibit, our advisory committee was surprisingly inarticulate. I noticed a similar pattern in subsequent conversations with fellow Catholics for whom the exhibit also touched a nerve. Strength of reaction varied, but in each case there was an inability to say precisely what it was about Body Worlds that troubled, concerned, or in many cases offended. Reference was often made to particular aspects of Catholic belief and practice thought to constitute weighty if not insuperable objections to von Hagens' show. "Surely there is some official teaching, canon law, or Episcopal pronouncement that explicitly condemns the plastination of dead bodies?" (There isn't, at least not yet.) "Catholic burial custom, and belief in bodily resurrection argue against such a practice, right?" (No they don't, at least not necessarily.) In fact, the Catholic cult of the saints, especially the public display of relics, as well as papal approbation of cadaveric dissection during the Renaissance origins of scientific anatomy (Vesalius, etc.) would seem to argue that, of all religious traditions, Catholicism would, or at least should have the least reason to object to the "anatomical exhibition of real human bodies." Still, for many practicing Catholics, at least in my experience, there is something *wrong* or at the very least *odd* about Body Worlds. Precisely what, however, it is difficult to say.

The following essay is simply an attempt to articulate more clearly what is problematic about Body Worlds from a Catholic perspective. Since at the heart of Catholic identity is the distinctive liturgical action known as the "Eucharist," or more colloquially, the "Mass," my remarks will have the character of a "theo-liturgical" critique. Such a critique does not operate independently of either the more popular or educated criticisms of Body Worlds. In fact, it presupposes them. However, it also goes beyond, or rather sharpens them, by drawing attention to aspects of Body Worlds that might otherwise go unnoticed. For example, that Body Worlds constitutes something of a "liturgical" rite all its own. I take up this theo-liturgical task in greater detail in the final section of the paper. The section immediately following examines the possible reasons why criticism of Body Worlds tends towards the *moral*, i.e., to the alleged sources of von Hagens' bodies. My argument is that such an objection, while obviously important and legally dispositive, fails to capture some of the deeper concerns raised by Body Worlds, concerns that do indeed take us into theological waters. In order to chart that course, however, we must have a firmer grasp on the phenomenon that is Body Worlds itself. Hence in the next section I argue that von Hagens' project represents a deliberately ambiguous concatenation of aims (educational, aesthetic, moral, commercial) that, taken independently, have not (until now perhaps) justified the sort of technical manipulation of the dead that von Hagens practices. Taken together, they allow von Hagens to play one aim off of another in an endless game of cat-and-mouse. In the end, however, I argue with philosopher Megan Stern that for all its post-modern vibe, Body Worlds is ultimately emblematic of the hyper-rationalistic medical culture dominant in advanced industrial democracies, and as such re-enacts the familiar modern narrative of romance/repulsion with respect to our bodies. In the concluding section, then, I suggest, by means of a theo-liturgical critique, that such a narrative can only be overcome liturgically.

II.

Objections to Body Worlds frequently tend toward the moral, and in particular to the problematic origins of at least some of von Hagens' plastinates. Without denying the importance of this criticism — which even von Hagens acknowledges, at least indirectly — we can ask why and even whether such objections capture the fullness of what seems to trouble many about the exhibit. Let von Hagens' *Landesmann,* novelist Günther Grass, say more explicitly what is often implied by such criticism of Body Worlds, and the principle of "informed consent" on which such criticism rests. For Grass, von Hagens is the "new Mengele," referring to the notorious "Nazi doctor," Josef Mengele, who conducted gruesome medical experiments on the captive inmates of Auschwitz-Birkenau, as well as accepting "donated" body-parts from other concentration and death-camps.[7] The first principle of the Nuremberg Code, formulated in the wake of the 1946 war-crimes trial of many of Mengele's collaborators, though not Mengele himself (who evaded capture), states that "the voluntary consent of the human subject is absolutely essential."[8]

The language of informed consent, and the principle of individual autonomy on which it rests, have been made even more familiar by the emergence of the contemporary "bioethics" movement in the 1960s.[9] The one and perhaps only "Great Commandment" of current biomedical research and practice is that individual autonomy must *not* be violated, which explains why so many of the bioethical controversies that dominate the public square of political and cultural debate, from Tuskegee to Schiavo, are reduced to arguments about whether a particular technique or policy will violate someone's "liberty interests," and also why the many others that don't as directly engage such interests (embryonic stem-cell research, genetic enhancement) appear so intractable. The protection of such interests is obviously central to any liberal democratic order, and in raising this issue here I am in no way suggesting that we could or should do without it. On the other hand, a one-sided emphasis on such protection, together with an expansive concept of those liberties thought to merit protection, have had an impoverishing effect on public moral discourse, especially in the United States, where deeper philosophical and religious questions tend to be marginalized in favor of procedural questions (such as whether informed consent has been obtained), and thus substantive moral debate displaced by arguments over "due process" and "equal protection." Moral argument over Body Worlds, I would claim, fits into this pattern.

The question, then, is whether Body Worlds engages and thus compels us to consider constitutive goods besides individual autonomy, and whether there are other questions to ask of von Hagens than where and how he acquired his specimens. If public criticism (as well as personal experience) is any indication, it clearly does and there clearly are. We might call the good that must be considered in addition to or beyond the good of individual liberty the good of the *human body* itself, and the question we ought to ask is whether von Hagens' Body Worlds presents an intelligible articulation of the good of the human body. Art historian Michael Lewis, in a brief but perceptive *Commentary* essay, asks precisely these questions. He concludes that von Hagens' show presupposes a "ghoulishly" unintelligible moral universe, but unfortunately, as we shall note, his criticism ultimately returns to the familiar objections to the sources of the Body Worlds' bodies.[10]

"The way a body is treated aesthetically," writes Lewis, "is always an index of a society's understanding of itself." On the basis of this principle, Lewis quickly traces the history of representation of the human body in the Western tradition. From classical antiquity to the present, western artists have presented the human body as expressing, among other things, an "ideal *moral* good" (e.g., Michaelangelo's *David*, da Vinci's *Vitruvian Man*). Emphases changed as understandings of the body changed. For example, the "artistic challenge" of medieval Christian art, in contrast to the art of pagan antiquity, was to render the body "corporal but not carnal," thus the use of folding cloth and elongated figures to bring forward the body in spiritual ecstasy. Under the influence of modernity, this pagan, Christian, and even Renaissance idealism gave way to a reductive materialism in which the body, bereft of spirit or soul, is more often represented, as in the medical portraits of Thomas Eakins (1844–1916) as "all there is," and thus in terms that accentuate the body's disproportion, imperfection, and vulnerability. It is only a short step to the deliberately grotesque representations of the human body, "sullied and debased," in the work of contemporary artists such as Hannah Wilke, Karen Finley, Cindy Sherman, and Andres Serrano.

Lewis notes that Body Worlds might seem to fit rather nicely into this trajectory of the progressively "emancipated" body, with its gauntlet of flayed and dissembled plastinates, and perhaps even cleverly to retrieve the Renaissance idealism of such works as Michaelangelo's *David* at the same time. And yet, claims Lewis, the nagging moral questions about the sources of von Hagens' bodies have proven too much for European audiences (or have at least since *Der Spiegel* raised the issue in 2004), where "historically, the mixing of murder and medicine is ... a more familiar and more horrifying cultural theme."

(Body Worlds has not toured in Europe since 2004.)* Thus, despite von Hagens' claims to the contrary, Lewis finds Body Worlds to be without moral purpose whatsoever, and in this sense more troubling even than the "abject art" of Finley, Serrano, et al.: "Insofar as they made art that was intentionally repugnant, they at least consciously presupposed the capacity to be disgusted — which placed them in an intelligible moral universe. No such presupposition, and no such capacity, are apparent in the Day-Glo candy-colored universe of Gunther von Hagens, whose work is forever antiseptic and clean, and whose sensibility seems as efficient as a bullet at the base of a grinning skull."[11]

Rhetorically powerful, Lewis' conclusion is nevertheless philosophically weak, as it stops short of articulating the larger (and less tractable) issues that an engagement with Body Worlds raises, most importantly, whether von Hagens presents us with an incoherent or distorted *understanding* of the human body. Like all such criticisms of Body Worlds, it also ignores the fact that the moral question about the origins of the show's plastinates is logically distinct from the moral questions raised by their public display. Would the same criticisms (e.g., of moral unintelligibility) apply if there were trustworthy documentation, say from an independent Institutional Review Board, that all the bodies in Body Worlds were those of individuals who had freely donated them to von Hagens? As mentioned above, von Hagens is highly sensitive to these objections about the origins of his plastinates, and has gone to great lengths (it would seem, not far enough) to address them. Whether and how he has addressed the larger issues that Lewis and others have only incompletely engaged, questions about the good of the human body, are other matters entirely.

III.

If one source of our inarticulacy over Body Worlds is the tendency in contemporary moral discourse to reduce substantive questions about constitutive moral goods to procedural questions about the protection of but one of them (individual liberty), another source is surely Body Worlds itself, which cleverly exploits both our natural and conventional ambivalence toward the human body, an ambivalence that religious doctrine and ritual have traditionally helped us understand and behave toward. As I will suggest, though, at the end of the day, the fascination and dread that Body Worlds so successfully evokes has less to do with our natural ambivalence toward the body, the form so skillfully manipulated by von Hagens, than with our current ambivalence toward the culturally dominant narrative about the human body displayed in various ways

**The exhibit returned to Europe with the premiere of Body Worlds 4 in Manchester, England, on February 22, 2008. See the Museum of Science and Industry at http://www.msim.org.uk/about-us/news/body-worlds-4.*

across contemporary medical practice. And in this latter sense, von Hagens and Body Worlds do not so much tell the story as exhibit it *par excellence*. And in this sense, as I will further suggest in the concluding section of the essay, Body Worlds can best be appreciated as a *parody* of the story about the human body articulated in Catholic Eucharistic practice.

It should be noted that von Hagens is keenly aware of the religious sensibility underlying much of the public reaction to Body Worlds. "Religious sentiments and concepts of morality, tradition and dignity," he writes, "have played a crucial role in shaping the way in which we handle corpses, and have given rise to a wide variety of burial rites in individual societies."[12] However, as with so much else about Body Worlds, the exhibit's actual engagement with and von Hagens' own considered judgments about this religious sensibility are highly ambiguous, purposely so it would seem. On the one hand, as someone trained in the science and art of human anatomy, and thus intimately familiar with the manipulation of "un-plastinated" dead bodies— the colorless, distended, putrefying sort — von Hagens expertly exploits the contrast between what we naturally expect (grotesque decomposition) and what we actually confront (odorless, stable plastinates) to its full and diverse effect. For some, the plastinates reveal what von Hagens refers to as "the beauty beneath the skin," what an older tradition of Renaissance humanism, which the exhibit conjures at great lengths, would have referred to as the nobility of the human form. For others, there is the experience of peering into the forbidding precincts of modern medicine. For still others the exhibit is a grim, though antiseptic reminder of human mortality, more pronounced precisely because of the relative permanence of the polymerized corpses. (In an unfortunate turn of phrase, at least for a post World War II German, von Hagens has claimed that each plastinate is capable of lasting for 1,000 years.) Of course, for many, Body Worlds is simply a chance to be safely entertained by the sight of "real" dead bodies, the same sort of fascination that draws people to roller coasters, horror flicks, and train wrecks. Indeed, one of von Hagens' longtime collaborators, Wilhelm Kriz, identifies this mechanism, the deliberate contrast between the revulsion we expect and the "plastic" we encounter, as one of the keys to grasping the popularity of Body Worlds.[13]

On the other hand, both Kriz and von Hagens interpret this experience as the breaking of a "taboo": "In our society, death is repressed, blocked out, so to speak and the corpses of other people, at least, are viewed with a revulsive shudder."[14] From this perspective, the eventual loss of this shudder by the end of the exhibit represents what Kriz calls a "personal victory" for the Body Worlds' visitor: "They [the visitors] had overcome the taboos that surround human corpses."[15]According to Kriz, it is this liberating effect of Body Worlds that also accounts for the strongest opposition to it (he mentions the controversy surrounding the Mannheim show in particular). Since, as Kriz also states, criticism has come chiefly from theologians, as well as pathologists and anatomists, it would be easy to interpret the controversy over Body Worlds as but the most recent chapter in what Andrew Dickson White famously called the "warfare of science with theology,"[16] but in this case with the advocates of science professing not to have debunked religious authority and belief but to have replaced it. "Whole-body plastinates," writes von Hagens, "join the ranks of skeletons and mummies as a new means of determining our post-mortem existence for ourselves."[17] Could attending Body Worlds, then, be meant as something like participating in a religious ritual or liturgy, at the conclusion of which one emerges with a new (and proper) appreciation of one's own bodily reality? The clear message at the end of the St. Paul exhibit was for the most part moralistic ("take better care of your body") rather than religious in any identifiable sense; however, quasi-religious themes were also evident: e.g., donate (i.e., make a gift of) your own body, and thus contribute materials, in turn, for the public spectacle of future generations who will likewise be awaiting their own education/emancipation.

How many Body Worlds visitors are conscious of the way in which the exhibit plays on our natural ambivalence toward the dead (and death), or its reoccupation of traditionally religious practices and doctrines is open to speculation. Probably very few. It is not coincidental that many of the shows' most vocal critics have come from the ranks of the academy, professional societies, and churches. Von Hagens' sleights of hand are rather familiar in these contexts (for one thing, there's too much money changing hands for the organizers to be entirely guileless). But one need not be a member of these relatively small guilds to sense that more is going on here than just a lucrative sideshow. Michael Lewis argued that Body Worlds presupposes an unintelligible moral order: the gratuitously grotesque, as it were. I would propose just the opposite; namely, that Body Worlds is all *too* intelligible, in that it represents, or in this case rather tangibly reenacts, the culturally dominant narrative about the human body on display in our current medical theory and practice. Within this narrative a set of constitutive goods jostles uneasily and often uncannily with each other: the good of health, of education, of aesthetic delight (beauty), technical skill, and profit. Alone, only the first two of these goods have traditionally justified our medical intrusions into and manipulation of the human body, especially the dead body. In Body Worlds, however, the dead human body is made to serve all of these ends (and more) simultaneously, which as some critics have observed, results less in its elevation than its reduction to a commodity. In this respect, Body Worlds mirrors our current ambivalence towards modern medicine, whose growing power to reshape our bodies both attracts and repels.

That Body Worlds aims to serve both a medical and educational good is obvious, as even the show's harshest critics have observed. As noted at the outset of this essay, plastination itself has its origins in a medical-educational milieux. As von Hagens' modification of Hippocrates' famous aphorism (*ars longa, vita brevis)* also reminds us: "Life is short, but Plastination will endure (and educate) for a long time."[18] Von Hagens and Body Worlds' promoters have proven adept at marshalling these goods in explaining and defending their exhibit. Indeed, the overwhelming message of the St. Paul show, like all the rest, was that in coming to understand the human body the visitor would leave with a greater appreciation for his own body, especially the importance of proper care for it. If for some reason that message is unclear, display cases full of diseased organs are there to drive home the point. One of the first "Gestalts" the visitor meets, at least in the St. Paul show, is *The Smoker*, whose visibly blackened lungs only reinforced this hygienic theme, which of course has become an enormously familiar one, at least for American (if not yet Western European or Asian) audiences. Here Body Worlds simply inserts itself, with all the authority of the medical establishment, into the ongoing war against big tobacco. Surely the fight to eliminate the public health menace of smoking justifies the display of a few plastinates?

In serving *public* health education, however, von Hagens is motivated by more than a desire to improve average life-span, and here Body Worlds' otherwise salutary pursuit of the goods of health and education is complicated by von Hagens' stated desire to liberate anatomy from its captivity to a traditional medical aristocracy. Journalist Stuart Jeffries, interviewing von Hagens shortly before the opening of the London Body Worlds show in 2002, noted that "von Hagens sees himself as on a global mission to end the elitism of the medical profession which, he believes, has denied the lay public access to a better understanding of their own bodies."[19] Even here, though, von Hagens is less radical than he would seem to think, for in the North American context at any rate, professional elitism has long given way before such things as the "patient's rights" movement, Health Insurance Portability & Accountability Act of 1996 (HIPAA), and WebMD, to name only a few. When the issue of professional authority does arise, recourse is normally had to the doctor's possession of superior technical skill, something that is, ironically, championed by von Hagens himself, each plastinate, for example, taking more than 1500 hours to process, etc.

In addition to the goods of public health, education, and technical skill displayed by von Hagens' plastinates, there are the more problematic goods, at least in this case, of aesthetic enjoyment (beauty) and profit. By itself, neither end succeeds as a justification of the public display of plastinated human bodies, even if those bodies were donated according to the established canons of informed consent. (Although if the critical acclaim accorded to Damien Hirst's displays of butchered livestock and formaldehyde sharks is any indication, then we may not be too far from losing our ability to shudder even here.) A full answer to why such ends would not by themselves justify von Hagens' enterprise is beyond the scope of this essay; however, where profit is concerned we meet arguments similar to those against the buying and selling of human organs, and the general repugnance to trafficking in human flesh, living or dead. The same goods informing the canons of informed consent (individual liberty), at least on one influential reading of them, cannot be alienated, even by autonomous moral agents themselves, even after death. At any rate, it is difficult to imagine anyone voluntarily submitting him or herself to what would amount to a species of post-mortem prostitution.

Plastination for aesthetic enjoyment would seem liable to the same critique, in this case less a matter of prostitution than pornography, a "pornography of death" as Geoffrey Gorer famously described it.[20] Interestingly, from this perspective Body Worlds hearkens back to the American mortuary and funeral practices so pointedly criticized by Jessica Mitford in *An American Way of Death.* Just as those practices were highly refined, aesthetic ways of denying death, so too the plastinated, "candy-colored," odorless corpses of Body Worlds quickly allow one to forget the dismal reality that makes the spectacle possible. There is as well the arresting knowledge that the "raw materials" at the artist's disposal are, in this case, not quite raw enough. Regardless of how many hours von Hagens spends shaping and posing his "Gestalts," the harder work of composing the human form has already been done. In fact, the more von Hagens "plays around" with his bodies, the more does the sublime intentionality informing the human body come to light, and, in return, all the more do von Hagens' manipulations appear as precisely that: "playing around."

North American audiences have become familiar, though not always comfortable, with a medicine shaped by aesthetic (and other non-therapeutic) ends, so again, in serving the good of aesthetic delight Body Worlds in this instance does not so much introduce the viewer to a radically new appreciation of the human body as mirror what is an already recognizable phenomenon: the body molded for aesthetic ends by scientific medicine. Von Hagens, for his part, resists the title of artist, though his demurrals can hardly be taken seriously. There is, to begin with, the cultivated resemblance to the controversial German artist Joseph Beuys, also known for his black fedora. Furthermore, the Renaissance anatomical illustrations that von Hagens' Gestalt plastinates quite obviously "quote" were themselves composed and understood in at least partially aesthetic terms. Finally, von Hagens has "signed" a number of his plastinates, much in the way a sculptor would sign his or her creation.

The multiple goods disclosed in Body Worlds make the show extremely difficult to categorize (Medical exhibit? Carnival side-show? Art Gallery?), and at the same time allow von Hagens, in good postmodern form, to play one meaning off the other in an endless round, one in which he can even claim to, at points, disappear as an author/artist. Meanwhile ticket sales increase. And yet, I would propose that Body Worlds does not so much provide the enlightened commentary on our own culture as it reflects the deep ambivalence about the human body characteristic of our current industrialized medical system. In this sense, the show's medical-education theme is key to understanding its other effects. Visitors to Body Worlds are thus presented with an understanding of the human body that is, *pace* Lewis, not unintelligible, but all too intelligible, in that the understanding on display is the understanding dominant in contemporary medical culture, a view of the body as the site for the display of a variety and play of forces that give us a very real, but also a greatly exaggerated, sense of control in the face of the multiple assaults on the body that we identify collectively as "mortality." "Medicine" is increasingly the name for a set of commercialized technical skills that promise to save us from our mortality, and promise us lasting if not radically extended control over it. The plastinated-for-public-display of the human body thus turns out to be another medical sleight-of-hand, in this case slowing down the aging process "post mortem."

Philosopher Megan Stern, among others, has drawn attention to the way in which Body Worlds' confluence of ends actually serves the commodification (rather than the promised liberation) of the human body in our current industrialized medical culture. Her comments will allow us to move, in the final section of this paper, to a fuller *theological* articulation of the concerns expressed by Catholics. "The exhibition," writes Stern, "is shocking not simply because it goes against the grain of what is apparently acceptable. It also represents, in the most dramatic terms, the redefinition of the human body within consumer culture. In this sense it is eerily representative of current values." "Body worlds attracts attention, whether from supporters or detractors, because it highlights, in spectacular ways, the changes that consumer culture has wrought upon ideas of medicine, the body and death. If it has been a recurring source of offence, hostility, legal intervention and allegation, this is perhaps because it articulates so clearly what is too close to home."[21]

IV.

The liturgy of the Eucharist, or as it is more colloquially known among Catholics, the "Mass," is neither a mechanism for moral instruction nor a means of social critique, at least not primarily. It is, rather, worship, which is its own proper form of activity, not to be confused with or reduced to other forms of human activity. Nevertheless, the power of the Eucharist (and of liturgy generally) both to form character and illuminate social and political realities has received increasing amounts of theological attention lately.[22] In this final section of the paper I draw on this body of work to help articulate what might be called a Catholic "theo-liturgical" critique of Body Worlds. Such a critique does not operate independently of either the common-sense reaction to or the more sophisticated philosophical and aesthetic criticisms of von Hagens' project. Rather, it presupposes them, expanding on and sharpening those reactions and criticisms by bringing forward elements within them that might otherwise go unnoticed—"To be trained in theology," James Gustafson once wrote, "should alert one to aspects of discussion that are otherwise hidden."[23] For starters, such an approach draws our attention to the liturgical character of Body Worlds itself. This is not a trivial point, for my argument, in a nutshell, is that Body Worlds is but a parody of the Eucharist, and as such constitutes an arresting inversion of the economy disclosed in Catholic liturgical practice.

"Economy," in Christian theological discourse, refers to the plan of divine salvation revealed in the person and mission of Jesus Christ (cf. Eph. 1:10), and thus to the mystery of the divine itself.[24] The term is essential for comprehending the nature of a theo-liturgical critique, for in juxtaposing Body Worlds and the Eucharist, more is being proposed than a sociological study of two seemingly disparate human performances. Both Body Worlds and the Eucharist, like all thoughtful human activity, present to their participants (who are never mere "on-lookers") the world perceived and intended in a particular way. In the Eucharist the world is disclosed as fundamentally *gift*: as the contingent creation of a radically generous Creator, whose own existence is neither completed by nor in any other way perfected by the act of creation. In the Eucharist the world is also *intended*, i.e., offered, as *gift*: as the only fitting return to the One by whose ceaseless giving the world simply *is*.[25] To celebrate the Eucharist is thus to enter into an "economy of gift," at whose heart lies the mystery of divine generosity itself. "Theology" is the reasoned exploration of this mystery by means of elemental distinctions and clarifications (e.g., eternal and finite being) that seek to render the mystery more intelligible, though not, for that reason, any less mysterious; as Sokolowski has written, theology involves a "raw and elementary kind of thinking," one that above all seeks to bring out the distinction between God and the world.[26] A "theo-liturgical" critique, therefore, is the reasoned investigation of temporal realities from within the setting opened up by the Eucharistic disclosure of divine love.

The human body is not incidental to the Eucharist: it figures centrally in the gifts of bread and wine that become, upon consecration, the body and blood of Jesus Christ, (who, in the words of Aquinas' famous hymn "*cibum turbae duodenae/se dat suis manibus*"*), but also in the body of the *ekklesia*, or assembled faithful, who both "make" the Eucharist and in so doing are also "made," i.e., into one body, by it.[27] "One bread, one body," goes a contemporary Catholic hymn, "one cup of blessing that is shared." From a Eucharistic perspective, then, the body too, like the world, is perceived and intended as gift. Indeed, the body—both the Eucharistic body of Christ and the body that is the Church—becomes the indispensable point of entry into the divine economy.

Like the Eucharist, Body Worlds invites its participants into a world in

*"*To his own apostles he gives himself as food with his own hands." The hymn is known as* Pange Lingua, *written by Thomas Aquinas (1225–1274) for the feast of Corpus Christi (the Body and Blood of Christ).*

which the human body figures centrally. Like the Eucharist, Body Worlds presents both a vision of the world and a vision of the human body in that world. Like the Eucharist, Body Worlds discloses an economy. In the economy disclosed by Body Worlds, however, neither the world nor the body is presented or intended as gift, or at least not unambiguously. As intimated in the preceding section of this essay, if there is a gift unambiguously on display in Body Worlds it is the gift of *technique*, which is to say, plastination, brought into high relief by its seeming ability to arrest bodily corruption indefinitely (from "ashes to ashes" to "ashes to plastic") and thus to overcome our natural revulsion in the face of it. It might be objected that the "Gestalt Plastinates" on display are indeed gifts inasmuch as they were presumptively donated with the free and full consent of the individuals they once embodied, but whatever gift-character they possess is far overshadowed by the technique that brings them to us as "plastinates." A contrast with Eucharistic liturgical action is instructive here.

For starters, the artfully flayed and mutilated bodies (and body parts) on display offer no further clue as to the identity of the donors. Indeed, in contrast to the Catholic rite, in which the impersonal elements of bread and wine ("fruit of the vine and work of human hands"[28]) become the living body of Jesus Christ, von Hagens' plastinates have been scientifically and systematically stripped of their identities and rendered as impersonal "Gestalts." As such, their character as artifacts ("the work of human hands"), and not as living beings, as unique and irreplaceable individuals, predominates; however, given the confidentiality requirements on most consent forms, this may be an insuperable difficulty. Indeed, if the bodies in Body Worlds are intended as occasions to wonder at the "beauty beneath the skin" (von Hagens' oft-used term), they are also and more forcefully there as mute witnesses to von Hagens' considerable skill, his ability, by posing the plastinates "just so," to obscure for us the fact that these bodies are *inanimate*. Von Hagens' penchant for hyperactive "sporting" poses, e.g., *The Basketball Player*, *The Skateboarder*, *The Archer*, etc., is important in this respect. (The slightest asymmetry here would break the spell.) Finally, unlike the living bread of the Eucharistic, sacramentally consumed, the plastinated figures of Body Worlds are inedible. They cannot be or become "spiritual food and drink," though they are nevertheless presented as items for "mass" consumption (as are the "relics" and "icons," i.e., tee-shirts, ball-caps and key-fobs emblazoned with images of plastinates that may be purchased in the exhibit shop). In this case, the Body Worlds' bodies are more akin to the toys in a fast-food "kid's meal," which like von Hagens' plastinates are also, ironically, mostly "made in China." One may be forgiven for wondering whether it all recycles.

As these remarks begin to indicate, then, from a theo-liturgical perspective, the economy celebrated by Body Worlds may more fittingly be described as an economy of death, but not death as such. Indeed, one does not need to be Catholic, Christian, or even liturgically-minded, to understand that the unique *frisson* deliberately cultivated by Body Worlds is the experience of staring death, so to speak, in the face. What a Catholic theo-liturgical perspective brings forward is the character of the death displayed in Body Worlds as "technically-mastered" death. Of course, the Catholic Eucharistic is also a celebration of the overcoming of death; not, however, through technique (there is a startling absence of technique in Catholic liturgy, which is part of what distinguishes it from magic), but through the ritual opening of a horizon in which the transcendent source of all life is disclosed in the unique and irreplaceable body of Christ, grasped, touched, tasted, and adored as such in faith.

From this setting, then, the economy disclosed in Body Worlds becomes recognizable as largely continuous with, and in fact hardly intelligible apart from, the economy on display in the industrialized, scientific medicine increasingly experienced as daily reality by denizens of the "developed" world. Like that medicine, Body Worlds both attracts and repels. It attracts in its apparent ability to spare us the indignities of suffering, death, and corruption; and to refashion the human form, perhaps endlessly, in a variety of aesthetically pleasant ways. But it simultaneously repels as we recognize the ironic cost of such freedom and pleasure is submission to an impersonal and ultimately arbitrary power — the power of technique — whose authority over us resides solely in its ability to conceal as long as possible its own helplessness before death.[29] The dying patient, inaccessibly sequestered in the intensive care unit, surrounded by intrusive life-prolonging machinery, visually captures the body-obscuring and ultimately body-denying tendencies of contemporary medicine. Body Worlds has given us a polymerized version.

In conclusion, a personal *coda*. As a participant in the decidedly modern *leiturgia* that is Body Worlds (invited to a special sneak preview of the exhibit, I was spared the costly admission-price), I too was alternately fascinated and repulsed by the seemingly endless assortment of plastinates. Well before the final display, however, I found myself growing increasingly bored by it all. Neither morally outraged nor aesthetically offended, but bored ("the banality of technique?") and far more intrigued by the living souls taking it all in than the plasticized vestiges of von Hagens' imagination. Such were and are the reactions of one Catholic. To the obvious, and trivial, objection that not every Catholic is likely to have the same (negative) reaction as mine, I can only reply that it may indeed be true, despite Church teaching, that bad liturgy can hurt you.

Notes

1. von Hagens, "Anatomy and Plastination," 25.
2. Ibid., 26.
3. von Hagens, "On Gruesome Corpses," 260.
4. von Hagens, "Anatomy and Plastination," 33.

5. Institute for Plastination, Gunther von Hagens' Body Worlds.
6. Paterson, "Body Worlds impresario."
7. Darwent, "In the Flesh."
8 U.S. Government Printing Office, *Trials of War Criminals*, 2: 181–2.
9. Jonson, *The Birth of Bioethics*, passim.
10. Lewis, "Body and Soul."
11. Ibid.
12. von Hagens, "On Gruesome Corpses," 279.
13. von Hagens, "Anatomy and Plastination," 6; see also "On Gruesome Corpses," 272–4.
14. Kriz, "Foreword," 6.
15. Ibid.
16. White, *A History of the Warfare*, passim.
17. von Hagens, "Anatomy and Plastination," 34.
18 Ibid., 35.
19. Jeffries, "The Naked and the Dead."
20. Gorer, *Death, Grief, and Mourning*, 192–9.
21. Stern, "Shiny Happy People."
22. See especially Shuman and Volck, *Reclaiming the Body*; see also Cavanaugh, *Torture and Eucharist*; Rubin, *Corpus Christi*.
23. Gustafson.
24. Cf. "Ökonomie (I)" in Kaspar, *Lexikon*, 7:1014f.
25. On the Christian concept of God as distinct from the world, see Sokolowski, *The God of Faith and Reason*. On the Eucharist as a presentation of this distinction see, idem., *Eucharistic Presence*.
26. Sokolowski, *The God of Faith*, 27.
27. The phrase "The Eucharist Makes the Church" has become associated with the work of twentieth-century Catholic theologian Henri deLubac. See his *Catholicism*, 88–111.
28. *Sacramentary*, 371.
29. For further elaboration of modern medicine as the "god that failed," see Hauerwas, *God, Medicine, and Suffering*.

Bibliography

Cavanaugh, William T. *Torture and Eucharist: Theology, Politics, and the Body of Christ*. New York: Blackwell, 1998.

Darwent, Charles. "In the Flesh," *The Independent Online*, December 2, 2001. http://findarticles.com/p/articles/mi_qn4158/is_20011202/ai_n14439544/print (accessed November 19, 2007).

Dickson White, Andrew. *A History of the Warfare of Science and Theology in Christendom*. New York: D. Appleton, 1896.

Gorer, Goeffrey. *Death, Grief, and Mourning*. Garden City, NY: Doubleday, 1965.

Hauerwas, Stanley. *God, Medicine, and Suffering*. Grand Rapids, MI: Wm. B. Eerdmans, 1994.

Jeffries, Stuart. "The Naked and the Dead." *Guardian Unlimited* (Online), March 19, 2002. http://www.guardian.co.uk/g2/story/0,,669775,00.html (accessed November 17, 2007).

Jonson, Albert. *The Birth of Bioethics*. New York: Oxford University Press, 1998.

Kaspar, Walter, Ed. *Lexikon für Theologie und Kirche*. Freiburg in Breisgau: Herder, 1993–2001.

Kriz, Wilhelm. "Foreword." In Whalley, *Gunther von Hagens'*, 5–8.

Lammers, Steven, and Allen Verhey. *On Moral Medicine: Theological Perspectives in Medical Ethics*. 2d ed. Grand Rapids: Eerdmans, 1998.

Lewis, Michael. "Body and Soul." *Commentary Online*, January 2007. http://www.commentarymagazine.com/viewArticle.cfm/Body-and-Soul-10814 (accessed November 12, 2007).

Lubac, Henri de. *Catholicism: Christ and the Common Destiny of Man*. San Francisco: Ignatius Press, 1988.

Paterson, Tony. "Body Worlds impresario 'used corpses of executed prisoners for exhibition.'" *Telegraph.co.uk*, January 25, 2004. http://www.telegraph.co.uk/news/main.jhtml?xml=/ news/2004/01/25/wbody25.xml&sSheet=/news/2004/01/25/ixworld.html (accessed November 15, 2007).

Rubin, Miri. *Corpus Christi: The Eucharist in Late Medieval Culture*. Cambridge: Cambridge University Press, 1991.

Sacramentary (The Roman Missal, 1970). New York: Catholic Book Publishing, 1985.

Shuman, Joel, and Brian Volck. *Reclaiming the Body: Christians and the Faithful Use of Modern Medicine*. Grand Rapids, MI: Brazos Press, 2006.

Sokolowski, Robert. *The God of Faith and Reason: Foundations of Christian Theology*. Notre Dame: University of Notre Dame Press, 1982.

_____. *Eucharistic Presence: A Study in the Theology of Disclosure*. Washington, D.C.: Catholic University Press, 1993.

Stern, Megan. "Shiny Happy People." *Radical Philosophy* 118 (2003). http://www.radicalphilosophy.com/default.asp?channel_id=2187&editorial_id=11213 (accessed November 16, 2007).

U.S. Government Printing Office. *Trials of War Criminals before the Nuremberg Military Tribunals under Control Council Law No. 10*. Volume 2. Washington, D.C.:, 1949.

von Hagens, Gunther. "Anatomy and Plastination." In Whalley, *Gunther von Hagens'*, 9–36.

_____. "On Gruesome Corpses, Gestalt Plastinates and Mandatory Interment." In Whalley, *Gunther von Hagens'*, 260–82.

Whalley, Angelina, ed. *Gunther von Hagens' Body Worlds: The Anatomical Exhibition of Real Human Bodies*. Translated by Francis Kelly. Heidelberg: Arts & Sciences Verlagsgesellschaft, 2005.

Twilife: the Art and Science of Consuming Death

Lucia Tanassi

Death is an inevitable part of life. To manage our curious discomfort with death, we have devised a powerful clinical apparatus of denial including forms of entertainment that allow us to get close to death in a non-threatening way. Plastination is a recent corollary to the American worldview on death. It is a complex process that allows the long-term preservation of human tissues by replacing body fluids with polymers (silicone, polyester, etc.), thereby stopping organic decay. The process gives resilience and consistency to tissues, producing single organs, organ configurations, or full body specimens that are dry and odorless and that can last for an indefinite period of time. The versatility of plastinated anatomical specimens has taken the plastination technique from a strategy of preservation to facilitate scientific inquiry to a strategy of performance for public education and entertainment.

The uniqueness of plastination is that it imbues new performativity on "real" dead bodies, and it fosters a new cadaver experience in the hyper-reality of the science museum exhibit. Von Hagens' vision for plastination is to promote awareness about public health issues such as smoking or obesity, and the overwhelming response of educators who are flocking to the exhibitions with their students is a signal that consumers are responding to this new sociology of death. It must be acknowledged however that this new sociology of death is rooted in the concomitance of opportunities for dismemberment and performance, commodification and consumption. The new cadaver experience also sustains a biopolitical order that sets apart dead bodies from living ones, unclaimed bodies from donated ones, and human biomaterials from human beings.

Twilife

This essay proposes twilife as a new interpretive concept to explore the ambiguities embedded in the social context that human biomaterials inhabit, and the relations that they produce. Twilife is a concept based on what Giorgio

Agamben calls the "zone of exception."[1] Beginning with homo sacer, in the context of Roman jurisprudence, and then in the context of the Jewish holocaust, Agamben discusses how political life (bios) lies within and under the protection of state sovereignty, while bare life (zoe) lies within the "zone of exception"—a space beginning at the limits of state sovereignty where the rules that define and protect living beings no longer apply.[2] In the context of the holocaust the "zone of exception" enabled the extermination of millions of Jews effectively transformed by Nazi propaganda into pests, prisoners, and non-persons. In the context of anatomical science the "zone of exception" enables a transformative process that fosters the availability of human bodies and body parts.[3] Plastination crystallizes the transformation of human bodies from political entities (with rights and personhood) into anatomical specimens and exemplars of "bare life," and plastination exhibits make specimens available to the mass market, but the "zone of exception" is further enabled by the scientific enterprise, and is also under the protection of bioethics committees.

The Los Angeles debut of Body Worlds, and many subsequent exhibits, included an ethics review that added scientific legitimacy, and successfully displaced questions about the exhibitionary use of cadavers and the use of unclaimed bodies. An ethics advisory committee, gathered in anticipation of Body Worlds' debut at the Los Angeles California Science Center, asked internationally recognized bioethicist Hans Martin Sass to travel to Heidelberg and carry out an "independent review of the exhibit's body donor documentation."[4] It is unfortunate that this independent review was limited to donor documentation instead of extending to an actual specimen count, as well as an assessment of protocols and policies to handle bodies procured through the donor program and unclaimed bodies procured from "government agencies."[5] The ethical review summary instead relies on Dr. Angelina Whalley's confirmation that whole-body plastinates as well as organ plastinates are from body donors, but crucially, the summary and the committee do not provide clear and independent verification that all bodies used in the exhibits are from donors.[6]

Dr. Whalley, Dr. von Hagens' wife, and managing director of the Institute for Plastination (IFP), is hardly an independent stakeholder. Yet her assurances contained in the summary of the ethical review, and the review itself, became an established bioethics precedent for other science centers and their advisory committees. Several of the science centers merely reviewed the background work carried out for the Los Angeles exhibit instead of performing their own assessments of provenance and due diligence. This ran counter to one of the key recommendations made by the California Science Center Advisory Committee that for future plastination exhibits: "...a special review has to verify that bodies and organs have been donated with full and informed consent of the donors. To exhibit human bodies or organs without free and informed consent is not acceptable."[7] Yet some committee members did not find the need for further verification because they trusted the work of colleagues who had participated in the California Science Center Advisory Committee. Some committee members specifically mentioned that it was not necessary to "reinvent the wheel" because their colleagues had good reputations and because there wasn't enough funding for that process anyway.[8]

By recognizing "informed consent" as a determinant of acceptability, the advisory committee acknowledged the longstanding tradition of principles laid out in 1979 with the Belmont Report.[9] The Report was prepared by the National Commission for the Protection of Human Subjects of Biomedical and Behavioral Research, and its purpose was to identify basic ethical principles that should shape the conduct of biomedical and behavioral research involving human subjects. The report outlined a three-principle framework that included respect for person (narrowly defined as autonomy), beneficence, and justice. These three principles are still widely recognized in research ethics as the "standard of practice," and have also become important benchmarks in clinical ethics. In research or in clinical practice the principle of respect for person is often narrowly interpreted as individual autonomy, which is commonly, although often inadequately, safeguarded by the informed consent form — a widespread device of ethics compliance. This narrow interpretation of respect for person as autonomy, and the focus on compliance and the consent form, underscore the limits of the "ethics review" in mediating on behalf of vulnerable subjects. Moreover, it is problematic that the "ethics review" can be held up as a symbol of legitimacy in the public domain and in the scientific community. In this sense the ethics review does not help to safeguard vulnerable subjects (be they potential specimens, museumgoers or both), but it helps museums and promoters to carve that zone of exception where what was once morally objectionable becomes a scientific pursuit of inherent educational value. This transformative process is how twilife begins to unfold as a biopolitical process of dismemberment, performance, and consumption. With the unfolding of twilife, human remains are at once re-signified as anatomical specimens, personable cadavers, and consumables. This biopolitical process reveals power relations between people, places, and projects like the profane logistics of body procurement (dismemberment), the high objectives of the anatomy exhibit (performance), and the profitable promises of an educational-entertaining science exhibition (consumption).

Twilife is not a static notion, but rather a transformative experience or process enabled by the state of exception. To sketch some of its limits, it is a state, understood both as a zone with its own political boundaries and as a domain where the status of human beings/remains is reshaped. The morgue, for example, is one such zone where the authority of the coroner and the laws may designate a body as unclaimed, or a few shards of bone as precious remains for a loved one. Twilife is a mode of being that refers to the social experience of death, the mode of being of plastinated specimens, and the interpersonal cadaver experience museumgoers may have with plastinated human remains.

Finally, twilife is an attribute that refers to human bodies or body parts that, although no longer "living," continue to have a social life of their own as museum specimens or as donated organs imbued with important transformative potential.

Dismemberment

Plastination was developed in the 1970s by Gunther von Hagens, an anatomo-pathologist from the former East Germany, who established the Institute für Plastination (IfP) in Heidelberg Germany in 1994. An anatomo-pathologist with a penchant for controversy, von Hagens performed a televised public autopsy in London in 2002 that produced substantial interest and outrage.[10] Since 1999, the British public had been absorbed by the Alder Hey Hospital scandal about the retention of organs from children who had died in surgery. The children's organs were retrieved as a matter of routine for pathology examinations, but were never returned to the parents, who had no idea the organs had been removed.[11] In this climate, many were disturbed by Dr. von Hagens' dissection talents and thespian inclinations. Allegations that were already circulating against von Hagens on suspicion of his involvement in the traffic of human body parts grew further. Of particular interest were von Hagens' ties to former Eastern Block countries (Siberia, Kyrgyzstan) and the People's Republic of China (PRC) that allegedly resulted in the acquisition of the bodies of vulnerable individuals (prisoners, homeless, mentally ill) "sold" to von Hagens (through the IfP) as unclaimed remains.[12]

Von Hagens has had connections with institutions such as the von Hagens Plastination Co. Ltd., Dalian (People's Republic of China); the Institute of Bioplastination Products, Dalian Medical University; the Plastination Research Centre at the Kyrgyzstan State Medical Academy, Bishkek (Kyrgyzstan); and the Novosibirsk Medical Academy (Siberia). Since 2004 he has been a visiting professor at the New York University College of Dentistry where he is designing the first anatomy curriculum based on plastinated specimens instead of cadaver dissection.[13] After the 2004 debut at the Los Angeles' California Science Center, Body Worlds found renewed success at Chicago's Museum of Science and Industry, Philadelphia's Franklin Science Museum, Toronto's Ontario Science Center and many other sites across North America.

Human remains destined for plastination are either processed at the IfP in Heidelberg (Germany), or the processing site in Dalian (PRC).[14] In January 2004 the German magazine *Der Spiegel* published a report by Sven Röbel and Andreas Wassermann alleging that von Hagens was involved in plastinating the remains of former Chinese prisoners.[15] Röbel and Wassermann's report, which won the Henri Nannen Prize for "best investigative achievement," alleged that von Hagens and his associates had sought opportunities to acquire "fresh bodies" and body parts by dealing with Chinese government agents. They also reported that at the Dalian facility some bodies bore signs of bullet head-wounds (a trademark sign of execution). Röbel and Wassermann also questioned the moral and financial transparency of von Hagens' operations, and cited circumstantial evidence against von Hagens such as the presence of prison camps and a military restricted area around Dalian, and the existence of offshore accounts in Gibraltar in the name of "International Exhibition Ltd.," a firm allegedly linked to von Hagens. In March 2005, however, a temporary restraining order was issued against *Der Spiegel* prohibiting the publication or dissemination of statements indicating that Body Worlds used the remains of "Chinese execution victims" pending imprisonment or a fine of ‹250,000.[16]

Other allegations surfaced from Siberia to Kyrgyzstan based on revelations and legal proceedings against individuals who had been involved in the procurement of bodies processed at the IfP. Vladimir Novosyolov, Chief of the regional forensic medical examination bureau in Novosibirsk, was accused in 2001 of attempting to ship about fifty unclaimed corpses and about four hundred brains to the IFP in the fall of 2000.[17] Novosyolov allegedly ordered all city hospitals to send all cadavers that had not been autopsied or claimed by their relatives to the forensic lab, after which the human remains were sent to the Novosibirsk Medical Academy morgue that had signed an agreement for "scientific exchange" with the IfP. Under this agreement the Novosibirsk Medical Academy would ship the unclaimed remains to Germany as "unnamed goods," and would receive back the plastinated bodies to be used in training, but after the first shipment an investigation began.[18] Although acquitted at the first trial in 2003 and at the appeal in 2004, Novosyolov was found guilty on charges of abuse of authority and sentenced to pay a fine of 1,224.00 (USD) on June 17, 2005.[19]

In 2003 the *British Medical Journal* reported that von Hagens was under investigation for allegedly using bodies without consent in connection with the plastination research centre at the Kyrgyzstan State Medical Academy and that two hundred bodies of unexplained origin were found in his institute. He was accused by a Kyrgyz MP of having "abducted" several hundred bodies from former Soviet prisons, hospitals, and psychiatric asylums without consent.[20] Von Hagens admitted that he did not know whether the families had been informed. At the Kyrgyzstan State Medical Academy in Bishkek, Valery Gabitov (head of the surgery department) said that his department had been exchanging "staff and materials" with the IfP, and the human remains were sent from Kyrgyzstan to Germany after receiving three hundred thousand German Marks from von Hagens.[21] Suiutbek Nazarkulov, General Director of the Republican Centre for Mental Health (Bishkek, Kyrgyzstan), also said that "157 unclaimed bodies of people [who] died in the mental clinics were given to the Morphology Institute of KMA [Kyrgyzstan Medical Academy], according to an agreement between the academy and the Gunter von Hagens clinic, the corpses were sent to Germany."[22] Von Hagens continues to maintain that he was unaware of these situations and that he has since terminated those relationships.[23]

All these facts taken individually are peculiar, if morbid, gossip. Taken collectively, however, they point at a strategy of outsourcing that exploits gaps in the regulation and exchange of anatomical specimens within and across countries. Outsourcing body procurement is the ultimate strategy to bypass and deflect the responsibilities that come with the procurement process. In tandem with the "ethics review," outsourcing is another approach to carving out that zone of exception that makes possible the dismemberment and amalgamation of bodies that are either expropriated (prisoners), or that are already abandoned beyond the reach and protection of political power and can be appropriated as unclaimed bodies (inmates, the mentally ill, and poor).

Performance

Plastination and Body Worlds are practices of dismemberment that nourish our appetite for science while skillfully obscuring certain ethical and social concerns about the procurement of human bodies and body parts. Plastination is surely one of the last century's greatest innovations in Anatomy. It is reshaping the visual and moral language of anatomical representation, while also rediscovering and reconfiguring death and the cadaver in the context of the well-accepted, contemporary model of the science museum. Much like the comfort found in organ donation, this new "experience" may help individuals give new meaning to their existence, extending it with renewed personhood, and may help medical students and the lay public to understand the hidden wonders of the human body. All these fascinating new opportunities, however, should not come at the expense of presumed donors abandoned beyond the limits of legal tutelage by law and ethics processes alike.

The experience and attributes of the zone of exception that I called twilife are a bi-product of death, bureaucracy, power, and all the cultural work we do to give meaning to our existence. Plastination contributes substantially to this work. Plastinates share important characteristics with the *neomorts* imagined by Willard Gaylin,[24] or François Dagonet's *faux vivants*[25] defined by their "new" social function to enable society to benefit from medical/scientific progress in the transplant field thanks to the "invention" of brain-death. Plastinates enable a new educational (but also voyeuristic) process. Although clearly dead, human plastinates appear almost alive, stabilizing and normalizing the unique clinical experience of dissection and dismemberment. *The Fencer* and *The Chess Player* are invented identities for public amusement and consumption. These anatomical identities, invented ad-hoc by the plastinator, create a new embodied connection between cadaver-specimen and viewer, a connection that is established through the act of mirroring oneself in the dissected plastinated specimen. In the experience of twilife, the sight of plastinated skinless flesh or blood vessels generates awe, but the presence and sight of tattoos or body hair stand as reminders of the person that specimen once was. As one visitor told me, "It's really amazing, I got so close to the basketball player that I could see the hair of his eyebrows, the skin of his lips, and then I realized these are actual people!"

Plastination exhibits re-craft the natural and initiate a process of habituation to twilife. The hospital is a carefully crafted "natural" place for the sick; the morgue or the funeral home are "natural" places for cadavers; or the hospice is a "natural" place for those who, although not quite dead, are rapidly approaching the end of life. Plastination, like the hospital, the morgue, or the hospice, helps create and keep a safe distance between the public and the cadaver, death and decay. Plastination neutralizes organic decay, takes away bodily effluents and smells, and allows the visitor to have new kinds of encounters with death and disease. When stepping in the state of twilife, death and disease are at once revealed and manipulated, exposed and concealed — made palatable for the paying public. Plastination "sanitizes" death and the cadaver becomes an instrument of scientific inquiry and an object for public display.

Consumption

There is little regulation over the current use of unclaimed bodies by medical or other institutions, although recent scandals in tissue procurement have increased the public's attention to the commodification of bodies and body parts. State statutes regulate who has jurisdiction over unclaimed bodies. In Tennessee, for example, the medical examiner may consent to the procurement of organs and tissues for transplant, and may decide after a period of no less than 96 hrs (4 days) to distribute unclaimed bodies to medical, dental, and anthropologic institutions. Whether or how the state should safeguard the unclaimed remains of those who are often most vulnerable — homeless, mentally ill, undocumented citizens, or non-citizens— is rarely addressed. In the United States, for purposes of education or tissue procurement, this is an accepted standard of practice. Donors, patients, and the public can realign their moral compass to fit the needs of tissue procurement and those in need of tissue transplant by suspending consent to donation for individuals whose remains are unclaimed or for whom information on body disposition is lacking. What justifies this realignment are discourses of need, waste, salvage, and presumed gifting, but the underlying distinction between donated and unclaimed bodies that allows their taking is that they're invisible in life as much as they're hidden in death. Unclaimed remains often belong to undesirable people already dwelling in the "state" of twilife. What could be done in Los Angeles for the remains of mentally ill patients from Kyrgyzstan or Siberia? At least in the exhibit they can be useful. It does not require a great effort to realign the moral compass to fit the needs of museums, promoters, educators, and the public. This may even be acceptable, but it should not be obscured by a presumed ethics "due process."

Human bodies, donated or unclaimed, often move through a profitable global exchange of body parts.[26] The dialectics of benefit and exploitation produce and reflect social anxieties about the ways in which the human body is being mined, parceled out, and transformed into a valuable commodity. Tsuyoshi Awaya has argued that the global traffic in body parts is a new form of cannibalism,[27] while anthropologists Margaret Lock and Nancy Scheper-Hughes have written extensively about the "rapacious demands" and "commodifications" in organ procurement.[28] The advent of transplant medicine has given new scientific legitimacy to the exchange and sale of body parts, but various forms of medicinal cannibalism have fascinated the "civilized" West throughout history. Euro-Americans have tended to think of cannibalism as a form of savagism and don't like to think of this practice as their own. However, physicians through history have often prescribed drinking human blood, consuming human flesh, or preparing medicines from corpses.[29]

In Amazonia where funerary cannibalism is well-known, the practice was not focused on the act of consumption or flesh eating. The Wari,' for example, practiced a form of funerary cannibalism, because the idea of disappearing into the tribe members' bodies was more comforting than the thought of one's own flesh rotting into the ground. Eating bodies was a way to make them disappear so the family could come to terms with the loss; consumption of human flesh was driven by compassion for the family.[30] Western forms of contemporary cannibalism, with few exceptions, no longer include eating of the flesh; rather, consumption of human flesh takes place in highly specialized environments and is mediated by highly specialized practitioners (organ and tissue procurement specialists, transplant surgeons, or as in the present case, plastinators). While the Wari's focus on compassion made bodies disappear and relations among people endure, the Euro-American focus on consumption pushed the dead body into spaces of hyperreality where it was redesigned, manipulated, and improved before it could be viewed. Consumption, whether by sight, touch, or ingestion, is a pivotal moment in the dynamics of twilife, because it highlights the transformative properties of consumption and the biopolitical questions around the consumption of human bioproducts.

Organ and body donation processes attempt, although often fail, to efface personal relations and sever a person's ties with society so that the body (or body parts) may endure beyond death and save or educate another person. The body or body parts are compassionate "gifts" for critically ill recipients, "commodities" for organ procurement specialists or "educational tools" for scientists. The substantial difference between the Wari' and Euro-Americans, however, is that in the former case consumption is primarily driven by the needs of the deceased and their families, while in the latter largely by the needs of those on the receiving end of the exchange. In the same fashion, plastination efforts are centered on severing a person's ties with society and effacing a person's identity so that the body (or body parts) may be displayed with a new "anatomical identity." In plastination exhibitions, the act of consuming the body spans across several exchanges—donation, plastination, and exhibition—where the new anatomical identity is made public. Plastinated bodies and body parts trundle along in a moveable feast with mounting profits and no end in sight. We consume these bodies with our presence, our gaze, and our hunger for the experience. All such transactions, including the theft of the thirteen-week-old fetus from the Los Angeles exhibition, are at once individual and collective acts of consumption.

Closure

This is not an indictment against Dr. von Hagens, plastination, or Body Worlds; nor is this an endorsement. Rather, this is a reflection on the nature of our relationship with death, human demise, and the many ways we go about making and consuming bodies. The anatomical specimens in plastination exhibits create a new domain in which human remains can be observed—the state of twilife. The interpersonal cadaver experience of plastination exhibits gives the public a new mode of experiencing death—a way of being in twilife. The social life of the cadaver unfolds as the plastinates display to the public their personable new identities crafted by the plastinator—the attributes of twilife. These are crucial experiences, because they allow and even compel viewers to imagine themselves as pre-cadavers. In von Hagens' sociology of death, this would in turn induce viewers to reflect upon their own unhealthy habits, and ultimately upon their own mortality.

It remains to be seen whether plastination exhibitions will change our unhealthy habits towards death, such as fear, denial, or excessive healthcare spending. So far, for all its efforts to unveil the mysteries of anatomy and disease, plastination continues to successfully keep the death taboo under the thin veil of edu-tainment. In spite of von Hagens' message to "democratize anatomy," plastination exhibits convey that death and cadavers are and should be managed by specialists. Plastinators, anatomists, coroners, funeral directors, doctors, or forensic anthropologists remain the ones who are specially trained to interact with and practice on cadavers. These specialists can read and interpret signs of health, illness, or malice on the body; they re-create the body for presentation to the lay public. An embalmer would prepare a body for viewing at a funeral after expensive and complex (although often unnecessary) manipulations that usually aim at restoring the identity of the deceased. Plastination gets the body ready for public viewing by stripping the identity of the deceased and ascribing a new "anatomical" identity. The democratizing promise of plastination, to foster anatomical learning among the lay public by placing them in contact with anatomical specimens (like medical students learning, usually for the first time, about structures of the human body); or to make us better informed patients—partners in healthcare decision-making with our physicians—remains for now elusive.

Notes

1. Agamben, *Homo Sacer.*
2. Ibid.
3. Cohen, "Operability," 79–90.
4. California Science Center, "Summary of Ethical Review."
5. von Hagens, "Anatomy and Plastination," 30.
6. Tanassi, "Responsibility and Provenance," 36–38.
7. California Science Center, 6.
8. Tanassi.
9. Department of Health, Education, and Welfare, "The Belmont Report."

10 Singh, "Scientist or showman?" 468; Lee, "Over my dead body," s52–3; Tilstone, "Anatomy ... sensationalism or educational?" 76.

11. Bauchner and Vinci, "What have we learnt from the Alder Hey affair?" 309–10.
12. Harris and Connolly, "World trade in bodies"; Jones, "Use of Bequeathed and Unclaimed Bodies in the Dissecting Room," 102–7.
13. von Hagens, "Walking the Walk," 22–7.
14. Bohannon, "Gunther von Hagens," 1172–5; O'Keeffe, "Fury as corpse show comes to UK"; Paterson, "Body Worlds impresario."
15. Röbel and Wassermann, "Händler des Todes."
16. Institute for Plastination, "Press Releases & Statements 2006."
17. Voronov, "Court reviews 'Cadaver Case.'"
18. Munro, "Body 'Art' Exhibit Creates a Stir in Novosibirsk"; Rowland, "Scientist denies link to 'snatched' corpses."
19. Koshmarchuk, "Verdict on corpse smuggling upheld."
20. Singh, 468.
21. Munro.
22. Burke, "Kyrgyzstan: 19–25 October 2000"; Franchetti, "Hospital sold bodies of starved patients."
23. Ulaby, "Origins of Exhibited Cadavers Questioned."
24. Willard, "Harvesting the Dead," 23–30.
25. Dagognet, *La maîtrise du vivant.*
26. Scheper-Hughes, "The Global Traffic in Human Organs."
27. Awaya, "The theory of neo-cannibalism."
28. Lock, "Deadly disputes"; Scheper-Hughes, "The New Cannibalism," 14–7.
29. Conklin, *Consuming Grief*, 8–13.
30. Ibid.

Bibliography

Agamben, Giorgio. *Homo Sacer: Sovereign Power and Bare Life*. Stanford: Stanford University Press, 1998.

Awaya, Tsuyoshi. "The theory of neo-cannibalism" (in Japanese). *Japanese Journal of Philosophy* 3 (1994): 29–47.

Bauchner, Howard, and Robert Vinci. "What have we learnt from the Alder Hey affair? That monitoring physicians' performance is necessary to ensure good practice." *BMJ* 322, no. 7282 (2001): 309–10.

Bohannon, John. "Gunther von Hagens: Anatomy's Full Monty," *Science* 301, no. 5637 (2003): 1172–5.

Burke, Justin. "Kyrgyzstan: 19–25 October 2003, Review of the week main events: 5. Scandal about usage of corpses," *Kyrgyzstan Daily Digest*. http://www.eurasianet.org/resource/kyrgyzstan/hypermail/200310/0040.shtml (accessed Feb. 13, 2008).

California Science Center. "Summary of Ethical Review 2004/2005." Body Worlds: An Anatomical Exhibition of Real Human Bodies. Los Angeles: 2004 http://www.koerperwelten.de/Downloads/ethics_summary.pdf

Cohen, Lawrence. "Operability, Bioavailability and Exception." In *Global Assemblages: Technology, Politics, and Ethics as Anthropological Problems*, edited by Aihwa Ong and Stephen J. Collier, 79–90. Malden, MA: Blackwell Publishing, 2005.

Conklin, B. *Consuming Grief: compassionate cannibalism in an Amazonian society.* Austin: University of Texas Press, 2001.

Dagognet, François. *La maîtrise du vivant.* Paris: Hachette, 1988.

Department of Health, Education, and Welfare, "The Belmont Report: Ethical Principles and Guidelines for the Protection of Human Subjects of Research." The National Commission for the Protection of Human Subjects of Biomedical and Behavioral Research, April 18, 1979. http://ohsr.od.nih.gov/guidelines/belmont.html

Franchetti, Mark. "Hospital sold bodies of starved patients." *The Sunday Times*, February 13, 2005. http://www.timesonline.co.uk/article/0,,2089–1481620,00.html (accessed Feb. 13, 2008).

Harris, Paul and Kate Connolly. "World trade in bodies is linked to corpse art show." *The Observer*, Sunday, March 17, 2002. http://observer.guardian.co.uk/international/story/0,,668874,00.html (accessed Feb. 13, 2008).

Institute for Plastination. "Press and Release Statements 2006." Gunther von Hagens' Body Worlds: the Original Exhibition of Real Human Bodies. Heidelberg: 2006. http://www.koerperwelten.newmedianet.de/en/media/releases statements/releases statements 2006.html?edit — 030406.

Jones, D. Gareth. "Use of Bequeathed and Unclaimed Bodies in the Dissecting Room." *Clinical Anatomy* 7, no. 2 (1994): 102–7.

Koshmarchuk, Maxim. "Verdict on corpse smuggling upheld." *RIA Novosti* (Russian News & Information Agency), August 22, 2005 11:24. NOVOSIBIRSK. http://en.rian.ru/society/20050822/41200947.html (accessed Feb. 13, 2008).

Lee, John A. "Over my dead body." *Lancet* 364, Suppl. 1 (2004): s52–3.

Lock, Margaret. "Deadly disputes: understanding death in Europe, Japan, and North America." *Doreen B. Townsend Center Occasional Papers* 4, no. 725 (1996).

Munro, Robin. "Body 'Art' Exhibit Creates a Stir in Novosibirsk." *The St. Petersburg Times*, April 10, 2001. http://www.sptimes.ru/index.php?action_id=2&story_id=15133 (accessed Feb. 13, 2008).

O'Keeffe, Alice. "Fury as corpse show comes to UK: Rights group demands proof that the bodies are not those of political prisoners executed in China." *The Observer*, April 2, 2006.

Paterson, Tony. "Body Worlds impresario 'used corpses of executed prisoners for exhibition.'" *News Telegraph* (Filed: Jan. 25, 2004). http://www.telegraph.co.uk/news/main.jhtml?xml=/news/2004/01/25/wbody25.xml (accessed Feb. 13, 2008).

Röbel, Sven, and Andreas Wassermann. "Händler des Todes." *Der Spiegel*, Jan 19, 2004.

Rowland, Jacky. "Scientist denies link to 'snatched' corpses." *BBC News*, December 2, 2002. http://news.bbc.co.uk/1/hi/world/europe/2535471.stm (accessed Feb. 13, 2008).

Scheper-Hughes, Nancy. "The Global Traffic in Human Organs." *Current Anthropology* 41, no. 2 (2000): 191–224.

_____. "The New Cannibalism: A Report on the International Traffic in Human Organs." *New Internationalist* 300 (1998): 14–7.

Singh, Debashis. "Scientist or showman?" *BMJ* 326, no. 7387 (2003): 468.

Tanassi, Lucia M. "Responsibility and Provenance of Human Remains." *American Journal of Bioethics* 7, no. 4 (2007): 36–38. http://www.bioethics.net/journal/j_articles.php?aid=1220

Tilstone, Claire. "Anatomy ... sensationalism or educational?" *The Lancet* 6, no. 2 (2005): 76.
Ulaby, Neda. "Origins of Exhibited Cadavers Questioned." *National Public Radio, All Things Considered*, August 11, 2006. http://www.npr.org/templates/story/story.php?storyId=5637687 (accessed Feb. 13, 2008).
von Hagens, Gunther. "Anatomy and Plastination." *BODY WORLDS — The anatomical exhibition of real human bodies*, edited by Gunther von Hagens and Angelina Whalley, 9–36. Heidelberg: Institute für Plastination, 2002.
_____. "Walking the Walk: Recent Innovations in Dental Education at NYUCD, Life, Death, and One Man's Quest to Demystify the Inner Realms of the Human Body." *Global Health Nexus* 6, no. 2 (2004): 22–7. http://www.nyu.edu/dental/nexus/issues/fall2004/index.html
_____, and Angelina Whalley, eds. *Body Worlds — The Anatomical Exhibition of Real Human Bodies*. Heidelberg: Institut für Plastination, 2001.
Voronov, Konstantin. "We've Become Almost Like A Relatives [sic]: Court reviews 'Cadaver Case.'" *Kommersant* (Novosibirsk), Mar. 29, 2005. http://www.kommersant.com/page.asp?id=558500 (accessed Feb. 13, 2008).
Willard, Gaylin. "Harvesting the Dead." *Harpers*, September 1974, 23–30.

Emergent Bodies: Human, All Too Human, Posthuman

Ara Osterweil and David Baumflek

Between the mass public exhibition of dissected human cadavers in the traveling Body Worlds show, and the celebrated, hyper-realistic figurative sculptures by artist Ron Mueck, it is clear that contemporary "art" is charting new corporeal territory. Although some critics cite the distinction between science and art to distinguish the exhibition of plastinated corpses at scientific institutions from the exhibition of Mueck's sculptures at more traditional fine art venues, both genres can be described as corporeal spectacles of the highest order. And while these works undoubtedly diverge from each other, methodologically, artistically, and perhaps even ideologically, what distinguishes their shared territory from earlier corporeal traditions is ultimately what connects them in the collective unconscious.

This essay aims to interrogate the ontology and the politics of these equivalently shocking corporeal spectacles to comprehend what is at stake in, and what is so disconcerting about, these perverse millennial representations of the alternately surreal and hyper-real cadaver. For despite the fact that only *one* of these exhibitions utilizes actual human cadavers, the uncanny appeal of both Body Worlds and contemporary body sculpture can be seen as analogous to the mix of attraction and repulsion that one experiences at the sight of the dead.

Unlike Body Art of the 1960s and '70s, which may be regarded as its logical forebearer, this contemporary body of corporeal work presents the human form from a radically depoliticized point of view. For in their quest to meticulously document the corporeal mechanics of either the real cadaver or the latex simulacra, both Body Worlds and contemporary hyper-real sculpture aspire to a supposedly universal condition of humanity. As the following essay argues, this artificial sense of universality is diametrically opposed to the idiosyncratic and highly individualized expressions of personal subjectivity that characterized the work of Body Art pioneers.

Being Human: Body Art in the 1960s

In order to understand the important ontological differences between the Body Worlds endeavor and contemporary hyper-realist sculpture, it is necessary to situate these current phenomenon in their appropriate art historical traditions. In order to do so, we shall sketch out two parallel but related tendencies, both of which were animated by an urgency to represent the human body in innovative and explicit ways.

While Western Art has always been invested in the representation of the human figure, Body Artists of the 1960s and '70s interrogated the traditional separation of artist and art object. Rather than drawing, painting, or sculpting the body in their work, Body Artists used the body "not simply as the 'content' of the work, but also as canvas, brush, frame and platform."[1] As part of a generation of artists for whom formalist criteria were considered insufficiently political, Body Artists recognized the body as a privileged site of power that could be transformed into a tool with which to challenge the dominant ideological apparatus. The body "that had been used, usurped, abused, displayed, a body that had been cut, wounded, [and] dramatized" became, in the 1960s, "the body that struggles, rebels, and indicates the escape from the coercions of power."[2]

What art historian Hal Foster has described as the urgent "return of the real" in the postwar avant-garde was founded upon both the political necessity of real action and the philosophical rejection of the fiction of disembodied subjectivity. For Body Artists, to be human was not to *have* a body, but to *be* a body. Conceived as an implicit attack on Cartesian notions of the subject as split between mind and body, Body Art was profoundly Conceptual. Although artists associated often-relied-upon sensational shock tactics in order to trouble the conventional relationship between artist and viewer, Body Art was, first and foremost, a practice of *thinking* with, through, and about the body *as* body. For Body Artists, embodiment was not something to be recognized abstractly, but the framework for both artistic and political rebellion.

Linked to the radical Minimalist practice of presenting the "thing as itself," Body Artists used "the body itself" to contest the problematic inscription of subjectivity, gender, sexuality, race, and ethnicity on the corporeal apparatus. For many artists, the taboo act of asserting the body, in all of its messy, corporeal "effervescence," was a deliberate act of political resistance. During a time of intense political and social upheaval, their visceral assaults on themselves (and the viewer) served as bold reminders of the corporeal carnage that characterized political conflicts at home and abroad. In the tumultuous era of the '60s and '70s, the act of displaying the abject, pained, suffering bodies by artists such as Ana Mendieta, Chris Burden, Yoko Ono, and Paul McCarthy was a way of making visible war's anonymous casualties—at home and abroad.

Within this highly charged historical context, the simple act of sitting upon a stage and inviting audience members to cut away parts of one's clothing, as Yoko Ono did in *Cut Piece* (1964), was a complex act of political confrontation that revealed the sadism of an audience willing to participate in the public violation of a vulnerable Asian female body. Similarly, Chris Burden's infamous piece *Shoot* (1971), in which the artist had a friend shoot him in the arm with a .22 caliber rifle from a distance of fifteen feet, implicated its audience by revealing their failure to intervene in the corporeal abuse of another human being. Far from being a senseless act of self-mutilation or a shameless spectacle designed to titillate audiences, Burden's performance was an intrepid attempt to forge an empathic relationship to pain between audience and performer, and, by implication, spectator and victim. As Burden asked curator Paul Schimmel, "How do you know what it feels like to be shot if you don't get shot?"[3] To cast oneself in the role of victim, as Ono, Burden, and other artists did, was to visibly and publicly identify oneself with the cast-out, violated, injured, dehumanized Other in the hope of enabling viewers to meaningfully consider the body in pain.

Human, All Too Human: The Act of Seeing with One's Own Eyes

Within the medical profession, the tradition of public autopsy can be traced back at least to the Renaissance. Within the plastic arts, however, the public spectacle of the dead or dissected body is a much more recent and rarified phenomenon.* Like the taboo against touching or coming into contact with the dead, the "artistic" use of actual human corpses had been unthinkable — until recently. With the ever-shrinking distinction between public spectacle and private ritual, as well as the concurrent re-territorialization of the flesh by technology, this taboo seems to be rapidly approaching its expiration date.

Artistic use of the wounded, dying, or deceased body has become more pervasive since the corporeal experiments of the 1960s. Within the increasingly robust "genre" of self-mutilation that artists like Burden, Ono, and Mendieta "founded," we find the grotesque defacements of the Viennese Actionists, the video-recorded facial surgeries of performance artist Orlan, and the masochistic self-flagellations of performance artists such as Bob Flanagan, Gina Pane, and Mike Parr. On the other side of the spectrum of contemporary art, we find the sadistic work of British *enfant terrible* Damien Hirst, whose scandalous display of dissected sheep seemed, for an ephemeral moment, the true emblem of our controversial age of cloning.

**As Rosenblum has pointed out, the use of real body parts in art was already evident in the hyperrealist polychrome sculptures of the Spanish baroque period, which used real hair, blood stains, and glass tears "to contribute to the uncanny illusion of a living, and usually suffering, human presence." Mueck, 72.*

For their own respective cultural moments, each of these artists thrust the pained, disfigured body to the limits of representation. Yet even as the taboos concerning treatment of the dead are constantly encroached upon, displaying an authentic human corpse remains particularly problematic. For obvious reasons, the public dissection of human cadavers broaches a host of moral, legal, and aesthetic issues that the dismemberment of farm animals can (though should not necessarily) evade. It thus comes as no surprise that the most thorough investigations of the dead body have — until now — occurred within the realms of cinema and photography. As founder of *Cahiers du Cinéma* André Bazin has argued, it was the advent of these technologies of mechanical reproduction that liberated the plastic arts from their "mummy complex" — or the aesthetic burden of preserving or "embalming" man's bodily appearance in the face of inevitable death.[4]

A different impulse than embalming is at work in Stan Brakhage's 1971 avant-garde documentary, *The Act of Seeing with One's Own Eyes,* in which the legendary experimental filmmaker documented the post-mortem dissection of human beings in a Pittsburgh morgue. The clever title of Brakhage's film, which references the etymological transcription of the meaning of the word "autopsy," suggests the profound psychological impetus that animates the ostensibly scientific need to witness the internal secrets of the human body. By making the scientific examination of the corpse available to the lay audience, Brakhage's film lays bare the "personal necessity of coming into contact with the human corpse in order 'to see for oneself' the reality of the flesh dissected"[5] Twenty years later, Andres Serrano's large format photography series *The Morgue* (beginning 1992), which depicts close-up details of corpses, demonstrated a similar longing to witness the obscured complexities of the flesh. Even more extreme than either of these projects is John Duncan's necrophilic "performance" piece *Blind Date* (1980), in which the artist recorded himself having sexual intercourse with a female corpse that he had purchased in Mexico. Following this experience of "indescribable intense self-disgust," Duncan returned to California to have a vasectomy in order to ensure that "the last potent seed I had was spent in a cadaver." After the operation, Duncan arranged a showing of *Blind Date* in order to explain to the public "what can happen to men that are trained to ignore their emotions."[6]

It is within this highly charged and potentially incendiary vein of "forensic" Body Art that we would like to situate Body Worlds. Like Body Art, Body Worlds insists upon using the real body, rather than a representation or simulacra, as its primary material. And like Body Art, Body Worlds is particularly interested in the *extreme body*— in this case, the dead, diseased, splendidly colored, stupendously posed, and possibly even necrophilic body.

In various traveling shows, Body Worlds invites viewers to scrutinize the usually hidden organs of actual corpses, preserved and made visible through the techniques of plastination developed by German anatomist Gunter von

Hagens in 1977. Here, bodies are splayed open to reveal an intricate design of musculature and arterial passages normally hidden from view by the protective armature of the skin. In many of the exhibits, bodies are posed in a vigorous parody of animation. Smiling ecstatically, engaged in convoluted positions of sport or play, these cadavers suggest that they are not only *real* human specimens, but, ironically, extraordinarily vital ones. Mocking the limits of living flesh, many of these human sculptures pose in genuinely outrageous fashion — holding their own removed skin up for contemplation (*The Skin Man*, 1997) or displaying the hidden compartments of their body as if spilling from a wardrobe (*The Drawer Man*, 1999). While the Body Worlds' literature explains that these appalling poses reference scientific studies dating back to the Middles Ages and the Renaissance, they more closely resemble the exploitative, sensational theatrics of a freak show.

In the aggrandizing style of turn-of-the-century impresarios like Barnum and Bailey, Body Worlds claims to be nothing less than the "world's most successful traveling exhibition," boasting 25 million visitors in over 40 cities in Europe, Asia, and North America since its inception in 1995.[7] Its unprecedented popularity is undoubtedly due to its use of real bodies, for as the official Body Worlds website maintains, "the authenticity of the specimens on display is essential" to the success of the exhibit. "It would be impossible to convey this anatomical individuality [what is elsewhere described as the "interior face" of the body] with models, for a model is nothing more than an interpretation. All models look alike and are, essentially, simplified versions of the real thing."[8]

According to von Hagens, who first realized the path towards plastination with the revelatory discovery of a meat slicer in a butcher shop,[9] the display of real human bodies is indispensable. Though von Hagens insists upon the inimitable pedagogical value of using real bodies, one wonders whether the stated mission of creating a place of "enlightenment and contemplation" is best served by von Hagens' methods. If having your body sliced like bologna and force-impregnated with plastic is considered the "enlightened" treatment of a human being, then one might be tempted to inquire about what medieval options are available.

As Brakhage, Serrano, and Duncan's work demonstrate, the need to "penetrate" deceased flesh often exceeds the purely scientific desire to study the body's functions and dysfunctions. Though Body Worlds makes its interest in scientific pedagogy explicit, one wonders whether the spectacle of the human cadaver can *ever* be detached from our inherent voyeurism, curiosity, and prurience. Why does Body Worlds strive so assiduously to disavow the potential of its cadavers to elicit any response other than scientific enlightenment? Whereas Brakhage made his own not-purely-scientific interest in the autopsied body evident in his title (as well as his extremely personal, highly idiosyncratic style of filmmaking), Body Worlds insists upon two grand, but ultimately problematic, "truths": 1) the uncompromised, scientific objectivity of its endeavor, and 2) the singular genius of its intrepid creator.*

In our digitally manipulated, cinematically enhanced world, the historically privileged realm of vision has proved to be fundamentally unreliable. Within the context of increased surveillance and ever-decreasing individual privacy, Body Worlds seems a symptom of the simultaneous insatiability and insufficiency of vision rather than a resolution of it. Although Body Worlds emphasizes the didactic value of *seeing* the "truth" of the body's inner mechanics, its constant injunctions against touching the corpses suggests that *seeing alone* cannot satisfy the desires that Body Worlds incites. Like the long-standing Western prohibition against touching the corpse, Body Worlds' repeated prohibitions against touching call attention to audiences' barely repressed desires to *physically* rather than visually penetrate the bodies of the dead. Is it possible that the necroscopic desires that Body Worlds authorizes belie intensely necrophilic urges? Perhaps the desires that continue to make Body Worlds so popular are not so different, after all, from the scandalous urges that animated John Duncan's obscene performance.

Post Human? The Reinvention of the Body in Contemporary Figurative Sculpture

In the previous section, we situated Body Worlds in the tradition of Body Art in order to understand its particular insistence upon using real rather than simulated bodies. We would now like to turn to the work of contemporary sculptor Ron Mueck (born 1958). By putting Mueck's work in dialogue with Body Worlds, we shall elucidate the ways that these two contemporary phenomena distinctly fetishize bodies that exist on the border between the human and the machine.

The tradition of hyper-realism — of which Mueck is the latest and most exemplary incarnation — emerged in the early 1970s, on the heels of, and very much as a reaction against, the radical antics of 1960s Body Artists. Like Body Art, which departed from the non-objective work of the Abstract Expressionists, hyper-realistic sculptors were invested in the return of the body as subject.

**In the tradition of Conceptual artist Joseph Beuys, Gunther von Hagens claims the status of war victim in his fantastical biography. In a life defined by terrifying escapes, perhaps the most extraordinary adventure followed his birth in 1945 in a remote section of Poland, then part of Germany. As the Body Worlds website notes, "To escape the imminent and eventual Russian occupation of their homeland, his parents placed the five-day-old infant in a laundry basket and began a six month trek west by horse wagon." While no mention of the Holocaust is made, World War II is treated as a kind of historical primal scene, seconded only by von Hagens' two-year imprisonment in East Germany during the Cold War. A self-proclaimed victim of historical turmoil, political prejudice, misunderstanding, and childhood infirmity, the mythical origins of von Hagens' assumed genius are treated as an important part of the Body Worlds official saga.*

However, rather than using the real, often de-sanitized body as subject, method, *and* political tool, sculptors such as Duane Hanson and John DeAndrea labored to create convincing, but ultimately depoliticized, human facsimiles. Rather than using the raw material of the flesh — its fluids, its capacity for movement and sound, its ability to feel pleasure and pain — as tools, hyperrealist sculptors looked towards technology to reproduce bodies that were conspicuously non-effervescent.

Hyper-realistic sculptors, as they were designated in 1971, were not interested in displaying the abject or wounded body. In sharp contrast to Body Artists, hyper-realistic sculptors were resolutely *not* concerned with the body in pain, the victimized body, the dirty body, or even the non–Caucasian body. On the salacious side of the hyper-realist spectrum are John de Andrea's nudes which — like the airbrushed figures in *Playboy* magazine — aspire to, and potentially surpass, the classical (and classically misogynist) ideals of beauty, youth, and fitness that had animated much of traditional sculpture. Unlike Ana Mendieta's bloody rape victim or Paul McCarthy's obscenely defiled body, de Andrea's nudes were young, attractive middle-class Americans captured in exquisite and un-ironic poses of self-satisfied narcissism. More ethnographically inclined is the work of Duane Hanson, whose "family album of America" presents an anthropological gallery of middle class types such as typical tourists, waitresses, policemen, and shoppers.[10]

Like Pop Art from which it emerged, hyper-realism was invested in the synthetic face of America, and fittingly presented the human form as a commercial symbol of artificiality rather than an emotionally-charged, embodied subject. "More real than real Americans,"[11] hyper-realistic sculptures were, paradoxically, not real at all. Devoid of emotional content, the seeming immediacy of hyper-realism was a *trompe l'oeil*, an illusory effect of technology and technique. Diametrically opposed to the Conceptually-oriented Body Art, which strove to eliminate the mediating structure of representation by presenting the body as itself, hyper-realism celebrated representation as pure representation, "with the necessary elimination of the subject."[12]

While several contemporary artists, such as Charles Ray, Judy Fox, Reuven Cohen, Tim Noble, and Sue Webster, work in the contemporary genre of hyper-realistic figure sculpture, we shall look primarily at the work of Ron Mueck. For as art historian Robert Rosenblum has noted, within the "vast population of body doubles" that characterizes the facsimile body trend in contemporary sculpture, Mueck has pushed "the goal of human cloning farther than any of his colleagues, right down to each eyelash and toenail."[13] Indeed, as curator Susanna Greeves has acclaimed, "Mueck has created the most flawlessly hyper-real figures in art history, so effectively imitating nature that the categories of art, image and reality seem to be suspended."[14]

Indeed, since his debut in *Sensation: Young British Artists from the Saatchi Collection*, which opened at the Royal Academy of Arts in London in 1997, the work of the Australian artist has astonished both the public and the gatekeepers of the art world. Meticulously constructed from modern materials such as silicone and fiberglass, Mueck's sculptures are technically indebted to his work as a special effects model maker in the television, film, and advertising industries.

As Rosenblum has written, Mueck's contribution to *Sensation* "immediately established his singularity, not only in its fanatical insistence on the perfect replication of every hair, pore, wrinkle, or muscle that defines an individual face and body,"[15] but also in its impassioned emotional appeal to viewers. Approximately forty inches long, *Dead Dad* (1996–1997; see figure 5) is a miniature sculpture of the prone corpse of a nude, slightly elderly man, rendered in painstaking detail.

With a matter-of-fact pose to match its matter-of-fact title, *Dead Dad* depicts the most commonplace of all occurrences: the ordinary death of an aging man. As Rosenblum writes, "after all, this physical reality of death, the detritus of life, could always be seen in countless numbers at hospital morgues or funeral parlors."[16] Thus, what is extraordinary about the sculpture is then not so much what it represents, but the obsessive skill with which it is rendered. Here the impeccability of Mueck's craft — most striking in the hard-to-capture details of skin, hair, or nail — attempts to disguise itself as what philosopher Roland Barthes calls "punctum," or the capacity of a particular detail of an artwork to pop out and prick, bruise, or *move* the viewer with the recognition of the "wound" of the real.[17]

Is it genuinely the recognition of *the real* that is communicated through these ruptures of perception? Or is it the wizardry of special effects technology that masquerades as real pathos, as punctum? Though they may appear to blur the boundaries "between the natural and artificial" world, as Rosenblum claims,[18] there is nothing "real" or remotely human about Mueck's figures. Why, then, are these synthetic replicas so much more emotionally compelling to viewers than their human counterparts in the Body Worlds show? Why do viewers seem to experience Mueck's sculptures as *more real* than real?

For Barthes, the punctum could not be intentional. Unlike the "studium," which Barthes described as the sum of the artwork's cultural, anthropological, ethnological, and historical information, the punctum could not be "put there" by the camera operator, but could only occur accidentally, bursting through and puncturing the "studium" with the sudden force of its emotional impact. Yet as Mueck leaves no room for accident, his sculptures capture nothing spontaneously. Even with their surplus of anthropological data, Mueck's sculptures lack the accidental, idiosyncratic revelation of real emotion.

According to Barthes, the studium "is a contract arrived at between creators and consumers ... a kind of education ... which allows me to discover the *Operator*, to experience the intentions which establish and animate his practices."[19] Much of the seeming pathos in Mueck's sculptures is dependent upon

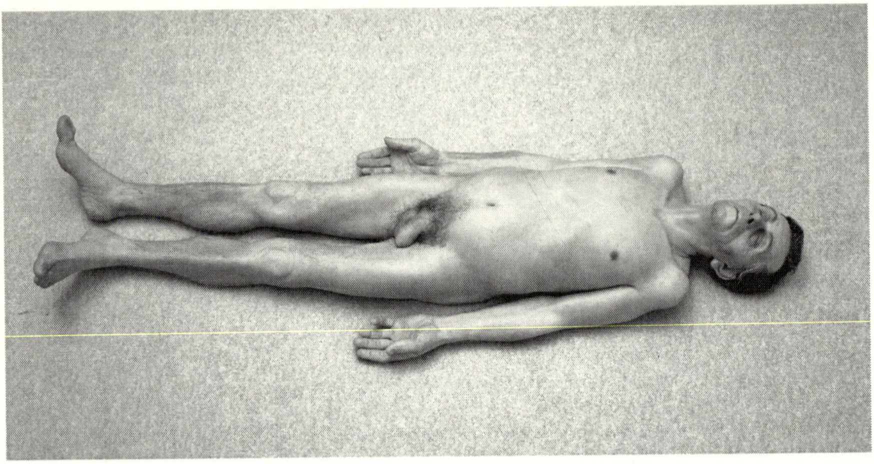

Figure 5: Approximately forty inches long, Ron Mueck's *Dead Dad* (1996–1997) acutely distills trauma's vocabulary. While its quotidian presentation of an ordinary corpse allows the viewer to inspect the nude male body without the niceties of funereal arrangement, the sculpture gains surplus emotional value from the knowledge that it depicts Mueck's own father. Courtesy and © Anthony d'Offay Gallery, London.

this contract between creator and consumer. Every detail bears the intentional mark of the technician whose painstaking process has created them. Like a true special effects operator, Mueck has endowed his work with what Barthes calls *functions*: "to inform, to represent, to surprise, to cause to signify, to provoke desire."[20] In a culture where images of the pained, traumatized body politic have been banished, Mueck's sculptures serve the perverse function of allowing spectators to feel empathy for plastic replicants. By relating so emotionally to Mueck's organless, fleshless "bodies," we purchase the right *not* to risk the emotional hazards of confronting the suffering of real human beings.

Bodies Without Organs Without Bodies

Certainly it can and will be argued that it is the vulnerable, poignant positions of Mueck's figures that move viewers to a state of startled cognition, in which the sensations of belief and disbelief coexist on a spectrum of emotion that includes empathy, intimidation, wonder, and delight. Indeed, many of Mueck's figures are posed in configurations pregnant with hopeful expectancy, anxious fear, quiet loneliness, or self-conscious timidity.

In order to counter this predictable objection, we shall briefly analyze three of Mueck's sculptures whose surplus of corporeal meaning has been scrupulously implanted by the artist in order to engage an *investigative* rather than *emotive* mode of address. When regarded together, these three sculptures

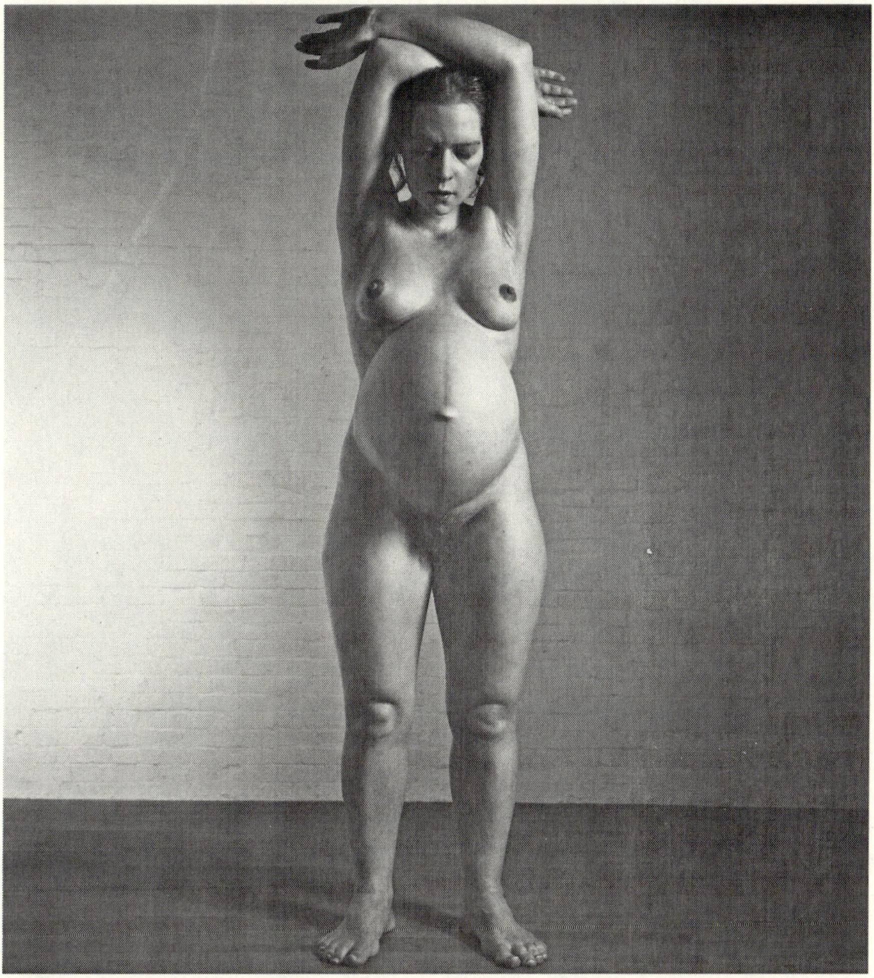

Figure 6: Mueck's larger than life *Pregnant Woman* (2002) leaves the viewer longing for X-ray vision in order to see the corpulent little fetus that undoubtedly resides inside her hugely distended stomach. Thus, the paradoxical fascination of the sculpture is contingent upon what presumably remains hidden within the (plastic) body's interior cavities. Courtesy and © Anthony d'Offay Gallery, London.

reveal a profoundly forensic mode of address that is comparable to the Body Worlds endeavor. We have already mentioned *Dead Dad*, whose quotidian postmortem pose allows the viewer to inspect the nude body of a man without the niceties of funereal presentation. To this explicitly forensic sculpture, we shall add two seemingly unlikely bedfellows: *Pregnant Woman* (2002) and *Mother and Child* (2001; see figure 7), both of which were constructed in conjunction with Mueck's residency at the National Gallery in London.

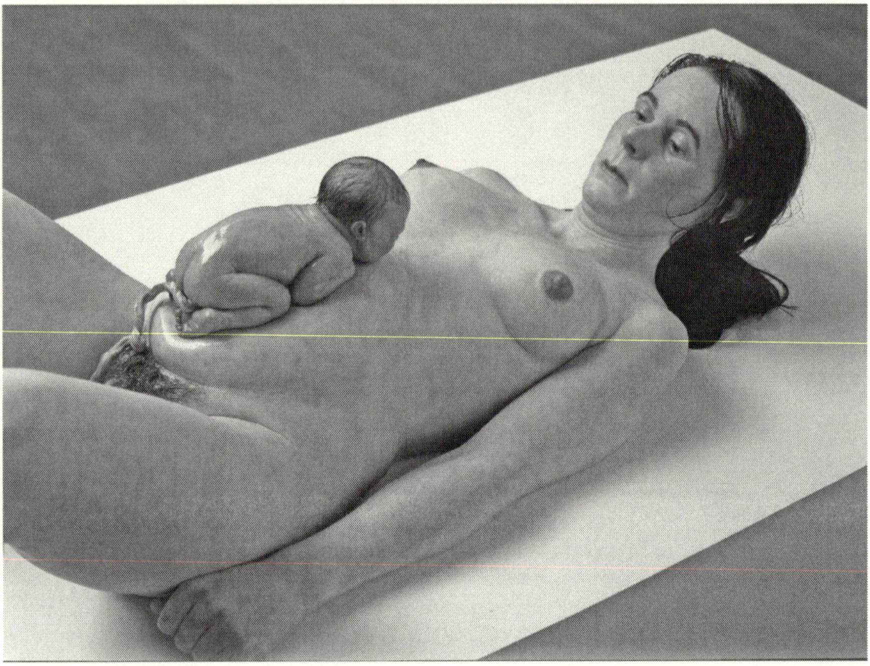

Figure 7: Mueck's minuscule, state-of-the-art *Mother and Child* (2001) insists that the fiberglass sculpture does indeed have real human insides, as the newborn seems to have emerged directly from Mother's womb rather than Mueck's hands. Yet like the classical Madonna figure she references, Mueck's Mother is also an empty vessel; her bewildered gaze seems to suggest that she alone knows the secret behind her own miraculous conception. Courtesy and © Anthony d'Offay Gallery, London.

Pregnant Woman is a larger-than-life portrait of an exhausted-looking woman, with stocky proportions and a hugely swelling belly, holding her arms above her head. Constructed entirely of fiberglass, the expectant, undeniably fleshy-looking mother looks as if her immensely swollen stomach is on the verge of bursting open to deliver what we imagine would be a corpulent little offspring.

Although it was completed a year earlier, *Mother and Child* seems to be a postscript to *Pregnant Woman*. This minuscule, state-of-the-art Madonna and Child depicts another exhausted mother. (Or is it the same one? Alas, it makes no difference, since, like von Hagens, Mueck is concerned with universal specimens, rather than particular individuals.) Here, Mother is presented presumably moments after she has delivered her mucus-slicked baby, who seems miraculously to have crawled upon her deflated stomach, umbilical cord still intact. As the babe gazes (demonically?) into the eyes of the sweat-dipped vessel from which he has emerged, the mother seems bewildered — if not terrified — by what has come out of her. Has she really birthed him? Has he really crawled from inside of her?

Perhaps it is not so much the newborn child that alarms Mother, but her own indeterminate state of being, since in spite of all of her apparently human characteristics, she has neither agency nor insides. In the classical Madonna tradition, the Virgin Mary (also an empty vessel) never seemed terrified of the precocious infant she had miraculously delivered. But Mueck's mother has ample reason to fear. The mystery of childbirth has remained a mystery to Mother, since she is, of course, completely hollow — and seems to know it. As in the story of the Immaculate Conception, it is the Invisible Patriarch —*and not Mother at all*— who has fathered these two overly-signified but ultimately empty beings into existence.

What unites these two sculptures, besides their obvious maternal theme, is their forensic or "autopsic" mode of address. Mueck's two Mommys (or shall we call them Mummies?) compel the spectator to imagine their internal organs. By making their fascination contingent upon what presumably remains hidden within the (plastic) body's interior cavities, Mueck dares us to see through their meticulously detailed fiberglass "skin." Longing for x-ray vision — or, better yet, a scalpel — we crave to examine their alternately jam-packed or just-emptied womb. The fact that their organs do not exist is, of course, both the point and the forensic foil of these works. Perhaps the strange glance exchanged between Mother and Child can be understood as a sly acknowledgement of this emptiness, this deceptive artificiality. As with *Dead Dad*, whose invisible cause of death must surely lie beneath his immaculate surface, the only way to penetrate the mystery of Mueck's mothers would be to slice them open.

Mueck enthusiasts are quick to defend his artistic integrity by suggesting that something beyond technology is at work. As Greeves insists, "Mueck's hyper-realism is not 'simply' the flawless technical imitation of reality, but requires another definition."[21] Yet, what Mueck's sculptures suggest is the paradox of a seemingly "humanist" art form that it is so thoroughly reliant upon technology that it implies, and perhaps even requires, the extinction of both traditional forms of figurative representation and traditional forms of human beings.

Nearly all of Mueck's admirers remark upon the almost tactile phenomenon of discovering the work — a significant paradox when one considers that they are made of plastic. As Greeves exclaims, "In the presence of a Mueck sculpture we are astonished by the perfection of the illusion: leaning in, it is impossible to fault, however close your range."[22] During our visits to Mueck's retrospective at the Brooklyn Museum in 2006, we were struck by the frequency with which security guards reprimanded viewers for their inappropriate proximity to the art objects, a practice that would not have been so remarkable had it not exactly mirrored our experience at Body Worlds. Even adults could not seem to help themselves from violating the socially acceptable distance of art viewing. In their insistence upon getting a closer look, viewers searched for telltale clues that would convince them, finally, that these tremendous plasticine clones were not human after all.

As in the Body Worlds exhibit, the urge to touch Mueck's sculptures is deliberately constructed to seem irresistible and universal. As Greeves writes, "The overwhelming desire to touch that *all* viewers seem to feel is an urge to corroborate their eyes' impression of living warmth and softness" (italics ours).[23] Again, the spectator's body must be disciplined to resist its almost involuntary urge towards palpation. Again, the mere act of *seeing* with one's own eyes seems inadequately stimulating or fulfilling. Like scolded children, we long to engage in the forbidden behavior of "seeing" with our hands.

Undoubtedly a result of his meticulous artistic process, Mueck's sculptures labor to persuade the viewer that they are made of flesh and blood. In spite of the prohibitions against tactility, Mueck's sculptures "ask" to be stroked; "fleshiness" is part of their studium. Not only can Mueck enable viewers to detect the subtlest trace of sweat on the brow of one of his silicone leviathans, but "he can also convince us that underneath their skins, his bodies must contain, like ours, fully defined internal systems of muscles, veins, and bones."[24] Often, it is the tiniest detail that makes the most persuasive claim to anatomical truth. Were it not for the flawlessness of these external details, we would never be compelled to insist, like Rosenblum, that any doctor "could give an anatomy lesson after cutting them open."[25] Of course, any doctor could *not* slice Mueck's bodies open and give an anatomy lesson since they are constructed of wire mesh, clay, fiberglass and silicone. These bodies, which are not really bodies, have no organs to speak of.

Disregarding the Pain of Others

> *No "we" should be taken for granted when the subject is looking at other people's pain.*
>
> — Susan Sontag, *Regarding the Pain of Others*

Unlike the Body Artists of the 1960s and '70s who used their own bodies as material — often endangering their own physical well-being in the process— the extensive team of scientists, medical doctors, designers, and technicians who are responsible for the manufacture of Body Worlds do not put their own bodies at risk. Rather than blurring the boundaries between subject and object as Body Artists did, here the traditional distinction between doctor and "patient," or in this case, medical specimen, is rigorously preserved.* "*We* are

**Since von Hagens' technique of plastination obviously requires the death of the specimen, it may seem ludicrous to critique the use of other people's bodies in his displays since one cannot reasonably imagine the roles of autopsee and autopseer collapsed into one. To be dead is to no longer be able to author, or so it would seem. And yet even this precious distinction has been challenged by Body (continued)*

doctors," the exhibit seems to whisper, attempting to disavow any suspicion that the bodies on display might have arrived there by less than scrupulous means. "Through the use of *our* technology," the hushed voice continues, "*we* have made these bodies available for *your* perusal." "But who are they?" one quietly wonders, to which the exhibit replies, "*They* are specimen; *their* identities do not matter."

In spite of an official rhetoric that superficially promotes the universality of the "medical specimens" on view in Body Worlds, the bodies on display are, by definition, radically Othered. Indeed, the very status of the unnamed, unclaimed, diseased, and impoverished corpses—disemboweled, flayed open, and cast out from their nation to be displayed for the "edu-tainment" of a paying audience—is irretrievably, irrevocably Other.

Considering typical Western attitudes toward death and the cadaver, the generally unperturbed demeanor of Body Worlds audiences is perplexing. Tracing modern antipathy to the corpse back to primitive societies, in which corpses signified the threat of contagion and the essence of violence, Surrealist philosopher Georges Bataille writes, "In the presence of the corpse, horror is immediate and inevitable and practically impossible to resist."[26] Likewise in *Powers of Horror*, contemporary philosopher Julia Kristeva describes the corpse as the penultimate symbol of abjection. For Kristeva, the sight of this "most sickening of all wastes" is so upsetting because it confronts the viewer with what he or she "permanently thrust[s] aside in order to live."[27] "A border that has encroached upon everything," the corpse is indeed "the border that has become an object."[28]

To our amazement, we detected no evidence of pathos, horror, disgust, or pity in the facial expressions, body language, or overheard conversations at the Body Worlds exhibit. Families gathered around the bodies and pointed, couples sniggered, children giggled. Flying in the face of the long-standing taboo in Western culture against looking at corpses, the mainstream popularity of Body Worlds suggests that perhaps traditional Western attitudes towards death have changed dramatically in the virtually post-human era in which we live. Unlike Shakespeare's Hamlet who contemplated Yorick's skull with a sudden awareness of his own mortality, viewers who examine von Hagens' plastinated corpses celebrate the triumph of technology over the organic body.

By making the autopsied, plastinated body available to the scientific and medical lay person, Body Worlds promises that each paying customer can approach and understand the mechanics of the formerly off-limits body with the expertise of a doctor. This discursive gesture rather misleadingly locates the spectator on the side of life and scientific mastery rather than on the side of

Artists. Consider performance artist Hannah Wilke, whose life-sized color photographs of her own body, deformed by cancer and medical treatment, were exhibited posthumously at the Ronald Feldman Gallery in New York a year after her death from lymphoma (Intra-Venus, 1993).

death, anonymity, and loss of corporeal control. We do not identify with the visible bodies on display (*those could not be us! we will never be like them!),* but with the invisible bodies (surely clad in white coats) who have placed them there for our perusal. As in Mueck's sculptures, where we ultimately identify with the Artist as Invisible Operator, in Body Worlds, we identify with the Artist as Invisible Anatomist rather than the bodies he has dissected.

Ontological Banishment and the Absence of Affect

With the Body Worlds exhibit, what is considered the reasonable treatment of human beings has plummeted. The corpses on display have been banished — not only from their home and nation — but, more importantly, from the very notion of what it means to be human in a civilized society. Apparently, their anonymity has reduced them to a kind of criminal or outlaw status, where their dead, naked, and dismembered bodies can be regarded utterly without pity by paying customers.

In his book *Homo Sacer: Sovereign Power and Bare Life*, Italian philosopher Giorgio Agamben describes the condition of the medieval bandit. Excommunicated from civilization, and ousted from the presumed sacredness of life, the bandit was considered "already" dead; it was thus legally permissible for *anyone* to kill him. Like Agamben's bandit, whose liminal existence was poised between life and death, the human beings in Body Worlds have been ousted from a civilized order in which the body is granted some degree of sacred capital. Yet as in Agamben's medieval scenario, the corpses displayed in Body Worlds are not completely outside of law and civilization, but a necessary component to it. While these corpses have been stripped of their own rights to participate in society and be protected by it, they, like Mueck's sculptures, continue to serve a crucial function for the civilized world.

Through its disavowal of pain and empathy and its radical de-individualization of the body, Body Worlds constructs the fiction of death without suffering. The illusion of living and dying painlessly — a hallmark of the opulent, leisure-obsessed society in which we live — is dependent upon the disappearance or banishment of pain, its forced non-existence. The notion that Other bodies exist to be sacrificed in order to enhance the leisure time of the paying customer is part and parcel of this cultural stance. As is consistent with capitalist ideology, we must purchase the experience of being in proximity to such profanity in order to disavow it. Like the law that pronounces the bandit "already dead" (and thus fit for sacrifice), the steep admission fee to Body Worlds guarantees our own immunity from death, mourning, and melancholia. For the price of admission, we are granted the privilege of not feeling. Or, in other words, by affixing our "first-hand" experience of death to a totem that can be left behind within the museum walls, we attempt to liberate ourselves from the burden of pathos. As we have purchased the right not to experience emotion for the doomed bandit, we refuse to bear responsibility for these outcast bodies.

In a world defined by global capital, "exile" no longer exists "out there" in the un-chartered territory beyond the city walls, but inside civilization, within the very institutions that traditionally defined it. By enshrining the dead, radically Othered body in a museum — the final resting place of all historic bandits and outlaws— Body Worlds has authorized both the disappearance of the human being and the banishment of death itself.

The Politics and Aesthetics of Trauma

> *We've never trumpeted so much crime against humanity now that science can no longer tell what is and what is not human.*
>
> — Sylvère Lotringer, *The Accident of Art*

Art history and theory may help illuminate the significant relationship between Body Worlds and contemporary hyper-realistic sculpture for our present cultural moment. In their book-length conversation *The Accident of Art*, theorists Sylvère Lotringer and Paul Virilio argue that a direct relation exists between trauma and art. Echoing Walter Benjamin's prescient discussion of the aestheticization of violence in "The Work of Art in the Age of Mechanical Reproduction," Lotringer suggestively declares, "Art is war by other means."[29] By characterizing contemporary art — in its trajectory from surrealism to terrorism — as a "victim" or "casualty of war," Lotringer and Virilio help us to understand the ways in which art of the twentieth and twenty-first centuries speak the trauma of the terrorized.

In the same way that you can't divorce painterly abstraction from its ontological roots in the disfiguration of the Second World War, one cannot approach either Body Worlds or contemporary hyper-realism without attempting to understand its larger significance for our cultural moment. What, in other words, is the relationship of both Body Worlds and contemporary hyperrealist sculpture to our current political, cultural, and historical landscape? What types of trauma do these disfigured bodies speak?

While both Mueck and von Hagens attempt to banish the experience of real trauma from the perceptible world (*all bodies are made of plastic, therefore none suffer*), Lotringer and Virilio suggest otherwise. For them, the trauma of contemporary art in general, and Body Worlds in particular, is inextricable from the defining anxiety of our time: the very definition of what it means to be human.

In his discussion of the Body Worlds exhibit that took place in Mannheim, Germany in 1997, Lotringer suggests that contemporary Body Art should be understood less as a "rediscovery of the body than a sort of farewell to any permanence it used to have."[30] Indeed, in a world where "fragmentation has become a new reality and the body a mere logo game," dead bodies "have taken over some of the attributes living bodies had to relinquish."[31] The "posthumous performance" of the dead in Body Worlds[32] may be a suitable return of the repressed for a culture that has forced the reality of death and suffering to the margins of consciousness. "Now that we no longer know what life is, or where it stops, we may feel the need to put death on display."[33]

In spite of its compelling insistence upon the "bodyness" of bodies, both Body Worlds and contemporary body sculpture replace the actual flesh and blood of human beings with plastic. "The paradox is that art over the last twenty years has tremendously emphasized the body, as though it had to show it one more time before it disappeared altogether. It wasn't a rediscovery, or a postmodern ressucitation [sic], it was post-mortem *before the fact*...."[34]

Rosenblum writes that by depicting "momentous events in the stages of life, love, and death," Mueck allows us to recognize his sculptures as "members of our own world."[35] What is so disconcerting about this claim in the context of the increasingly post-human world in which we find ourselves, is that *real* human beings— in Body Worlds and elsewhere — no longer seem to be recognizable as members of our own world. Why do we feel absolutely no pathos when confronted with evidence of the real organs that Mueck's sculptures lack? Why can we relate to human suffering better when it is rendered in plastic rather than flesh? It is as if our ability to identify ourselves with technology has risen, in inverse proportion, to our ability to identify with human beings.

Conclusion

In the context of the supposedly outstanding scientific and educational value of Body Worlds, moral indignation has been re-signified as an inappropriate, even cantankerous, response to the technological wizardry that disavows even the slightest moral ambiguity. In the postmodern, virtually post-human world that we precariously inhabit, the vocalization of humanist concerns has been rendered passé. And yet, these bodies are casualties of an unnamed war of scientific progress.

The rhetoric of the Body Worlds endeavor insistently locates the techniques of plastination in the realms of science and epistemology, suggesting of course that barbarism plays no role. While it would be comforting to think of science and barbarism as diametrically opposed systems, the history of science and technology suggest otherwise. Von Hagens' innovative use of technology does not neutralize questions of his enterprise's dubious ethics. On the contrary, considering the history of technology and the dangers of its fetishization, we should be even more sensitive to its potential abuses.

Our ongoing indifference to the casualties of contemporary warfare is mirrored by our indifference to the corpses on display in Body Worlds. When confronted with these human sculptures, we are not supposed to *feel* anything. That is, we are not supposed to feel anything akin to affect; in place of the recognition of pain, suffering, and loss, we are provided with a surplus of wonder, incredulity, and delight. Rather than experiencing empathy for the deceased individuals whose bodies constitute the display, we are invited to marvel at the scrupulous process by which these corporeal phenomenon have been so artfully exposed and composed.

Viewing the cadavers in Body Worlds is explicitly contingent on the suspension of structures of identification. For to actually believe in our shared humanity with the bodies on display is to admit to oneself the unspeakable spectacle of exploitation in which one has not only willingly participated, but paid a rather extravagant price to witness. If we *were* to contemplate the prior existence of these carcasses as once-vital individuals— each with unique and irreplaceable personalities— we would surely recoil in horror. For what we are being asked to contemplate, as we unthinkingly exchange curiosity for empathy, are the after-effects of a hyper-sanitized, scientifically approved *Grand Guignol* in which nameless dead bodies have been eviscerated for our viewing pleasure.

Body Worlds is an alarming milestone in the increasing de-personalization, de-humanization of the corporeal body. Unfortunately, so is the spectacular work of Ron Mueck.

Notes

1. Warr and Jones, *The Artist's Body*, 11.
2. Miglietti, *Extreme Bodies*, 15.
3. Burden, quoted in Warr and Jones, 122.
4. Bazin, "The Ontology," 9.
5. Osterweil, *Flesh Cinema*, 277
6. Duncan, quoted in Warr and Jones, 105.
7. Institute for Plastination, *Gunther von Hagens'*.
8. Ibid.
9. Ibid.
10. Miglietti, 213.
11. Ibid., 214.
12. Ibid., 215.
13. Rosenblum, "Ron Mueck's," 53, 54.
14. Greeves and Wiggins, "Ron Mueck," 43.
15. Rosenblum, 50.
16. Ibid., 46.
17. Barthes, *Camera Lucida: Reflections on Photography*, 27.
18. Rosenblum, 53.
19. Barthes, 28.
20. Ibid.
21. Greeves and Wiggins, 44.
22. Ibid., 43.
23. Ibid., 44.

24. Rosenblum, 54.
25. Ibid.
26. Bataille, *Eroticism*, 47.
27. Kristeva, *Powers of Horror*, 3.
28. Ibid., 269.
29. Lotringer and Virilio, *The Accident of Art*, 17.
30. Ibid., 51.
31. Ibid.
32. Ibid.
33. Ibid., 52.
34. Ibid., 32–33.
35. Rosenblum, 54.

Bibliography

Agamben, Giorgio. *Homo Sacer: Sovereign Power and Bare Life*. Trans. Daniel Heller-Roazen. Palo Alto, CA: Stanford University Press, 1998.

Barthes, Roland. *Camera Lucida: Reflections on Photography*. Trans. Richard Howard. New York: Hill and Wang, 1981.

Bataille, Georges. *Erotism: Death & Sensuality*. Trans. Mary Dalwood. San Francisco: City Lights, 1986.

Bazin, André. "The Ontology of the Photographic Image." In *What Is Cinema? Volume One*, edited and translated by Hugh Grayor, 9–16. Berkeley: University of California Press, 1967.

Benjamin, Walter. "The Work of Art in the Age of Mechanical Reproduction." Trans. Harry Zohn. In *Illuminations*, edited by Hannah Arendt, 217–51. New York: Schocken, 1968.

Burden, Chris. "Just the Facts." Quoted in Warr and Jones, 122.

Deleuze, Gilles, and Felix Guattari, *Anti-Oedipus: Capitalism and Schizophrenia*. Trans. Robert Hurley, Mark Seem, and Helen R. Lane. Minneapolis: University of Minnesota Press, 1983.

Foster, Hal. *The Return of the Real: The Avant-Garde at the End of Century*. Cambridge, MA: The MIT Press, 1996.

Greeves, Susanna, and Colin Wiggins. *Ron Mueck*. Exhibition Catalogue for show at National Gallery Company, London. New Haven: Yale University Press, 2003.

Institute for Plastination. "Prelude." Gunther von Hagens' Body Worlds: the Original Exhibition of Real Human Bodies. Heidelberg: 2006. http://www.bodyworlds.com./en/prelude.html

Kristeva, Julia. *Powers of Horror: An Essay in Abjection*. Trans. Leon S. Roudiez. New York: Columbia University Press, 1982.

Lotringer, Sylvère, and Paul Virilio. *The Accident of Art*. New York: Semiotext(e), 2005.

Miglietti, Francesca Alfano. *Extreme Bodies: The Use and Abuse of the Body in Art*. Milan: Skira, 2003.

Osterweil, Ara. "Flesh Cinema: The Corporeal Avant-Garde, 1959–1979." PhD diss., University of California, Berkeley, 2005.

Rosenblum, Robert. "Robert Rosenblum on body doubles." *ArtForum*, Oct. 2004.

_____. "Ron Mueck's Bodies and Souls." *Ron Mueck*. Exhibition Catalogue. Fondation Cartier pour l'art contemporain. New York: Thames and Hudson, 2006.

Sontag, Susan. *Regarding the Pain of Others*. New York: Picador, 2003.

Warr, Tracey, and Amelia Jones. *The Artist's Body*. London: Phaidon, 2000.

About the Contributors

David Baumflek is a Brooklyn-born sculptor whose work has been included in group and solo shows in New York, Philadelphia, and Boston, among other cities. With a background in political philosophy and architecture, Baumflek is currently pursuing advanced studies in the field of Art History. Baumflek holds an MFA from the Pratt Institute in Brooklyn.

Aviva Briefel is associate professor of English at Bowdoin College, where she teaches courses in Victorian literature and in film. She is the author of *The Deceivers: Art Forgery and Identity in the Nineteenth Century* (Cornell University Press, 2006) and is currently working on a new project on fin-de-siècle representations of the colonial hand. She is also co-editing an anthology of essays on the horror film after 9/11.

Christian DuComb is a PhD candidate in theatre and performance studies at Brown University. In addition to Body Worlds, his research interests include the Philadelphia Mummers Parade and the contemporary practice of classical Indian theatre. His essay "Present-Day Kutiyattam: G. Venu's Radical and Reactionary Sanskrit Theatre" appeared in the fall 2007 issue of *TDR: The Drama Review*. A theatre practitioner as well as a scholar, Christian creates and performs original dance/theatre with Dead Genius Productions.

Ruth Levy Guyer is the author of *Baby at Risk: The Uncertain Legacies of Medical Miracles for Babies, Families, and Society* (Capital Books, 2006). She teaches bioethics courses at Haverford College (visiting professor since 2002) and courses in the writing program at Johns Hopkins University (since 1994). She is a regular commentator on National Public Radio's weekend *All Things Considered* and is author of numerous articles, essays, and commentaries on bioethical, medical, and scientific topics. Her PhD is in immunology.

T. Christine Jespersen is a professor of English at Western State College of Colorado. Currently, she is researching literatures of globalization and environmental justice. She holds a PhD from Rutgers University, where she specialized in late nineteenth and early twentieth century American literatures and gender studies.

Stephen Johnson is director of the Graduate Centre for Study of Drama at the University of Toronto, where he teaches performance theory and the history of popular performance. His publications include the book *Roof Gardens of Broadway Theatres*, and articles in a range of journals, including *The Drama Review*, *Cana-*

dian Theatre Review, *Theatre Topics*, and *Nineteenth Century Theatre*, as well as *Theatre Research in Canada*, which he (co)edited for ten years.

Uli Linke is a cultural anthropologist at the Rochester Institute of Technology in Rochester, New York. Her research specialties include the cultural politics of memory, the visual culture of violence, and the political anthropology of the body. Her current projects include *Culture of Fear: The Intimacies of Global Violence* (forthcoming).

Natalia Lizama is an associate lecturer in the School of Social and Cultural Studies at the University of Western Australia. Her research interests include body theory, critical theory, posthumanism, medicine and biomedical technologies. She is currently working on a project that examines the relationship between the human and the animal in contemporary literature.

Natalie Loveless is a PhD candidate in the History of Consciousness program at the University of California, Santa Cruz. She is also a visual artist, a lecturer at the San Francisco Art Institute, and an independent curator/arts organizer.

Alexandra Ludewig is currently associate dean (education) and convener of German at the University of Western Australia. She studied in Germany, England, South Africa, and Australia and holds a doctorate from LMU Munich and a PhD from the University of Queensland. Since 1997 she has lectured in German and European Studies at the University of Western Australia and has published widely on Heimat and identity.

Peter M. McIsaac is assistant professor of German studies at York University in Toronto. His publications include *Museums of the Mind: German Modernity and the Dynamics of Collecting* (Penn State Press, 2007), and, as co-editor, a special issue of *New German Critique* on contemporary German literature (2003). His articles have appeared in *The German Quarterly*, *Monatshefte*, *Literatur für Leser*, and *German Life and Letters*. His current project is a book on German antecedents to Body Worlds.

Lisa Nevárez is an associate professor of English at Siena College. She holds a PhD in comparative literature from Vanderbilt University. Her research interests include Romanticism and the Gothic, as well as Latino/a literature.

Rebecca Onion is a graduate student in the department of American Studies at the University of Texas at Austin. Her interests lie in the sometimes-intersecting fields of environmental studies, technology studies, and the body. She is currently planning a dissertation that will examine questions of how children's cultures related to evolutionary thought and developing technology in the late nineteenth and early twentieth centuries.

Ara Osterweil is an assistant professor of film studies and art history at Muhlenberg College in Allentown, Pennsylvania. She has a PhD from the University of

California, Berkeley, where she wrote her dissertation on corporeal avant-garde cinema from the 1960s and 70s. Osterweil is a frequent contributor to *Film Quarterly* and has published extensively about sexually explicit cinema, most recently in *Porn Studies*, ed. Linda Williams, and *Women's Experimental Cinema*, ed. Robin Blaetz.

Patricia Pierson holds a doctorate in comparative literature from the University of California, Irvine, where she is assistant director of the Literary Journalism Program. Her research and teaching interests include collection theory, the contemporary novel, psychoanalysis, and life writing. Her current book project is an exploration of the idea of the author in relation to literary and pop-cultural representations of the undead.

Pedro Ponce teaches fiction writing, contemporary American literature, and literary theory at St. Lawrence University. He recently contributed an essay on the work of Mary Caponegro for *Narratives of Community: Women's Short Story Sequences* (Cambridge Scholars Publishing, 2008). His fiction and criticism have appeared in *Ploughshares, The Beacon Best of 2001, The Review of Contemporary Fiction, Rain Taxi, Small Spiral Notebook*, and other publications.

Alicita Rodríguez was most recently a Visiting Professor at Western State College of Colorado. Currently, she is taking a year off to complete a novel. Her creative and scholarly work have appeared in various journals, including *TriQuarterly, Denver Quarterly*, and *Review of Contemporary Fiction*. She is the Editor of the literary journal *Marginalia*.

Elizabeth Simon Ruchti is an interdisciplinary scholar whose research employs a performance studies methodology in her work in gender studies and queer theory. She has also published on film, religion, and performances of masculinity. Her current scholarship posits "queering" as a performance technique and a method of performance criticism in which the normative is de-centered from a queer perspective. Dr. Ruchti currently teaches theatre and gender studies at West Chester University of Pennsylvania.

Joseph Starr is an independent scholar, fiction writer, and translator. Recently, his work has been published in *Denver Quarterly, Mississippi Review*, and *Phoebe*. He is the poetry editor of *Marginalia*.

Lucia Tanassi is assistant professor of medical ethics with the Center for Biomedical Ethics and Society at Vanderbilt University Medical Center, Nashville, and she is a visiting scholar with the Department of Anthropology at the University of California, Berkeley. She works on broadly defined end-of-life issues from palliative care to organ donation, and has published on plastination in the *American Journal of Bioethics* and the *American Journal of Public Health*.

Paul Wojda (PhD, Christian ethics, University of Notre Dame, 1993) is associate professor in the Theology Department of the University of St. Thomas, Saint Paul,

Minnesota, where he specializes in biomedical ethics. He is also director of the MA degree program in Catholic studies, which promotes the scholarly reflection on and advancement of the Catholic intellectual tradition through the interdisciplinary study of its origins, development, and contemporary reception.

Index